1BDK3260

Variational Methods
in
Nonlinear Analysis

Variational Methods in Nonlinear Analysis

Edited by

Antonio Ambrosetti
Scuola Normale Superiore, Pisa, Italy

and

K C Chang
Department of Mathematics, Beijing University, China

Gordon and Breach Publishers

Australia • Austria • Belgium • China • France • Germany • India • Japan • Malaysia • Netherlands • Russia • Singapore • Switzerland • Thailand • United Kingdom • United States

Copyright © 1995 by OPA (Overseas Publishing Association) Amsterdam B.V.
Published under license by Gordon and Breach Science Publishers S.A.

All rights reserved.

No part of this book may be reproduced or utilized in any form or by any means, electronic or mechanical, including photocopying and recording, or by any information storage or retrieval system, without permission in writing from the publisher. Printed in Singapore.

Gordon and Breach Science Publishers S.A.
Postfach
4004 Basel
Switzerland

Library of Congress Cataloging-in-Publication Data

Variational methods in nonlinear analysis / edited by Antonio
  Ambrosetti and K.C. Chang.
     p.   cm.
  Includes bibliographical references and index.
  ISBN 2-88124-937-X
  1. Calculus of variations.   2. Nonlinear functional analysis.
I. Ambrosetti, A. (Antonio)   II. Chang, K. C., 1958–
QA315.V37   1993
515'.7—dc20   93-45010
                                                               CIP

Contents

Preface	vii
Geodesics On Space–Time Manifolds: A Variational Approach *Vieri Benci and Donato Fortunato*	1
The Role Of The Mean Curvature In Some Nonlinear Problem With Critical Exponent *Adimurthi, Filomena Pacella and S.L. Yadava*	39
Nontrivial Solutions To Some Nonlinear Equations Via Minimization *D. Arcoya and L. Boccardo*	49
A Morse Theoretic Approach To The Prescribing Gaussian Curvature Problem *Kung Ching Chang and Jia Quan Liu*	55
Evolution Problem Of Yang–Mills Flow Over 4-Dimensional Manifold *Chen Yun-Mei and Shen Chun-Li*	63
Periodic Solutions For A Class Of Singular Nonautonomous Second Order System In A Potential Well *V. Coti Zelati, Li Shujie and Wu Shaoping*	67
Nonergodic Properties Of Systems Of Oscillators Coupled Through Convex Even Potentials *G.F. Dell'Antonio, B. D'Onofrio and I. Ekeland*	77
Periodic Solutions Of Second Order Autonomous Systems: Existence And Multiplicity For The Prescribed Period Problem *Mario Girardi and Michele Matzeu*	83
Existence And Multiplicity Results Of An Elliptic Equation In Unbounded Domains *Massimo Grossi*	91
Multi-Bump Solutions To Scalar Curvature Equations On S^3, S^4 And Related Problems *Yanyan Li*	103
Geometry And Topology Of The Boundary In Some Semilinear Neumann Problems *Giovanni Mancini*	113

Closed Orbits With Prescribed Energy For Singular Conservative Systems With Strongly Attractive Potential 119
Lorenzo Pisani

On A Model Problem of Multiple Pendulum Type 129
Paul H. Rabinowitz

Perturbation Of Symmetry And Generalized Mountain Pass Theorems 137
Bernhard Ruf

A Two Points Boundary Value Problem On Non Complete Riemannian Manifolds 149
A. Salvatore

Multibump Solutions And Topological Entropy 161
Eric Séré

Avoiding Collisions In Singular Potential Problems 173
Enrico Serra

Periodic Solutions Of Hamiltonian Systems 187
Michael Struwe

A Note On Palais–Smale Condition And Mountain-Pass Principle For Locally Lipschitz Functionals 193
Stepan Terisan

Nonlinear Neumann Problems With Critical Exponent In Symmetrical Domains 205
Zhi-Qiang Wang

Some Results On The Geodesic Connectedness Of Lorentzian Manifolds 221
Antonio Masiello

Multiple Periodic Solutions To Some N-Body Type Problems Via A Collision Index 245
Pietro Majer and Susanna Terracini

The Role Of The Domain Shape On The Existence And Multiplicity Of Positive Solutions Of Some Elliptic Nonlinear Problems 263
Giovanna Cerami

Index 287

Preface

This volume collects most of the papers presented during the 14th course on "Variational Methods in Nonlinear Analysis" held at Erice, Sicily, from 12 to 20 May 1992, in the frame of the International School of Mathematics "G. Stampacchia".

The aim of the course was to discuss recent advances in the Calculus of Variations in the Large and in its applications to Nonlinear Analysis. The course was structured with plenary addresses on the state of the art, invited lectures and other shorter communications.

The meeting was attended by about 60 participants (from ten countries) whose discussions created a very stimulating atmosphere.

The course was sponsored by the Italian Ministry of Education, the Italian Ministry for University and Scientific Research, the Italian National Group for Functional Analysis and Applications (GNAFA–CNR), the Italian National Research Council, the Sicilian Regional Government, and the World Laboratory of Lausanne.

We wish to thank all these institutions for their generous financial support.

We owe special thanks to Professor A. Zichichi, Director of the Centre, for substantial assistance and to Professor F. Giannessi, Director of the School of Mathematics, for his boundless advice.

We also thank the personnel of the Ettore Majorana Centre for Scientific Culture who greatly contributed to the success of the course.

A. Ambrosetti
K. C. Chang

GEODESICS ON SPACE-TIME MANIFOLDS: A VARIATIONAL APPROACH

VIERI BENCI
Ist. Mat. Appl. "U.Dini"
56127, Pisa, Italy

DONATO FORTUNATO
Dip. Matematica-Universita'
70125, Bari, Italy

0. INTRODUCTION

Until the beginning of the 19th century the nature of the space as described by the Euclidean Geometry was considered an "absolute" truth.

The discovery of non Euclidean geometries provide the first evidence that this point of view is too limitative. After, at the beginning of the XXth century, the theory of General Relativity provided a "physical" evidence that the space, in the presence of matter, is not euclidean.

Moreover in General Relativity the space and the time are structurally joined: in this theory a gravitational field is described by assigning on a smooth 4-dimensional manifold M a metric tensor (which we denote by $\langle\,,\,\rangle_L$) having signature +++−. This means that, for all z in M, $\langle\,,\,\rangle_L$ is a non degenerate symmetric bilinear form (on the tangent space $T_z M$ to M at z) whose matrix representation has exactly one negative eigenvalue.

Such a metric $\langle\,,\,\rangle_L$ is called *Lorentzian metric* and M equipped with $\langle\,,\,\rangle_L$ is called *Lorentzian manifold* or *space-time*. Any point in M is called *event*.[1]

---

[1] Although the definitions and the results refer to four-dimensional manifolds, they can be extended in an obvious way to n-dimensional manifolds, n > 1.

The geometry and the topology of the space-time $(M, <,>_L)$ are determined by the Einstein equations which connect the Ricci curvature to the matter distribution.

This paper is devoted to the study of geodesics in space-times.

More precisely we are interested in three problems:

I) To find analytical and topological assumptions on a space-time M which guarantee the existence and the multiplicity of geodesics joining two <u>arbitrarily</u> given events in M (see sections 2,3,4,5).

II) To construct a Morse theory for the Lorentzian geodesics, i.e. to relate the "topology" of the path space on M to the geodesics joining two arbitrarily given points in M (section 6).

III) To develop an existence theory for the *periodic trajectories* i.e. for the Lorentzian geodesics whose "spatial components" are periodic (see section 7).

The aim of this paper is to give a short description of some recent results in these directions. To this end it will be useful to recall some basic facts on Lorentzian geometry (see e.g. [2, 28, 36, 37] for extensive treatments). We recall that a geodesic on M is a curve $z = z(s)$ solving the equation

$$D_s \dot{z}(s) = 0 \qquad (0.1)$$

where $D_s \dot{z}(s)$ denotes the covariant derivative of the velocity vector field $\dot{z}(s)$ along the direction of z. In other words the geodesics are the curves whose velocity vector field is paralleled translated along itself.

In terms of a local coordinates system with coordinates $z^1, \ldots, z^4$, a curve $z = z(s)$ determines four smooth real functions $z^1(s), \ldots, z^4(s)$. The equation (0.1) then takes the form

$$\ddot{z}^k + \sum_{i,j} \Gamma^k_{ij}(z^1, \ldots, z^4) \dot{z}^i \dot{z}^j = 0 \qquad (0.2)$$

where Γ^k_{ij} are smooth real functions on M (Christoffel symbols), depending on the metric tensor.

Equation (0.1) is the Euler-Lagrange equation related to the

"energy" functional
$$f(z) = \int_0^1 \langle \dot{z}(s), \dot{z}(s) \rangle_L \, ds \qquad (0.3)$$

Notice that the name "energy" we give to (0.3) is not related to the physical meaning of this functional.

If $z = z(s)$ is a geodesic, then the identity
$$\frac{d}{ds} \langle \dot{z}(s), \dot{z}(s) \rangle_L = 2 \langle D_s \dot{z}(s), \dot{z}(s) \rangle_L = 0$$
shows that $E = \langle \dot{z}(s), \dot{z}(s) \rangle_L$ is constant along $z(s)$. A geodesic $z = z(s)$ is called *spacelike, null* or *timelike* if E is greater, equal or less than zero, respectively.

In the theory of General Relativity, a timelike geodesic $z = z(s)$ is physically interpreted as the world line of a material particle under the action of a gravitational field. In this case the parameter s is called *proper time* and it is interpreted as the time measured by a clock associated with the particle. A null geodesic is the world line of a light ray. The spacelike geodesics have a more subtle interpretation: for a suitable local observer, they represent "Riemannian" geodesics consisting of simultaneous events.

Before to conclude this introduction we briefly indicate one of the main difficulties which occurs in developing an existence theory for geodesics in Lorentzian manifolds.

As already observed, the geodesics joining two given events z_0, z_1 in M are the critical points $z = z(s)$ ($s \in [0,1]$) of the energy functional (0.3) on the infinite dimensional manifold Γ^1 consisting of the smooth curves $z(s)$, $s \in [0,1]$, on M with $z(0)=z_0, z(1)=z_1$ (for a more precise definition of Γ^1 see (4.1),(4.2),(4.3)).

Unlike the situation for positive definite Riemannian metrics, whose geodesics can be searched by a direct minimization method, the functional (0.3) is unbounded both from below and from above, since the Lorentzian metric $\langle \, , \, \rangle_L$ is indefinite. Moreover it is not difficult to see that a critical point z of (0.3) is a saddle point with infinite Morse index (i.e. the Hessian form f"(z) of f at z possesses infinitely many positive and infinitely many negative eigenvalues). This fact does not permit a direct use of the usual critical point theories, because the "topology" of the sublevels of f does not change when a critical value is "crossed".

This paper is divided in eight sections:
1. Some examples of space-times.
2. Geodesics for R-complete static space-times.
3. Geodesics for static space-times with singular boundaries.
4. Outline of the proofs of theorems.
5. Geodesics for time-dependent space-times of split- type.
6. Morse theory for Lorentzian manifolds.
7. Periodic trajectories for static space-times.
8. Some other problems.

1. SOME EXAMPLES OF SPACE-TIMES.

1.1 Minkowski space-time

The Minkowski space-time is the manifold $M = \mathbb{R}^4$ equipped with the flat Lorentz metric

$$\langle \zeta, \zeta \rangle_L = \zeta_1^2 + \zeta_2^2 + \zeta_3^2 - \zeta_4^2 \,, \quad \zeta = (\zeta_1, \ldots, \zeta_4) \in \mathbb{R}^4 \tag{1.1}$$

This is the space-time of Special Relativity and it describes situations in which the gravitational effects are negligible.

Since the metric tensor (1.1) does not depend on $z \in \mathbb{R}^4$, equation (0.1) reduces to $\ddot{z} = 0$. Then the geodesics of the Minkowski space-time are just the straight lines of the underlying Euclidean space $\mathbb{R}^4$. The null geodesics through a given event $z \in \mathbb{R}^4$ form a cone with vertex z.

1.2 Static and stationary space-times

Consider $M = M_0 \times \mathbb{R}$, where M_0 is a smooth, connected 3-dimensional manifold on which we assign a smooth Riemannian metric $\langle \cdot, \cdot \rangle_R$ (i.e. a positive definite metric tensor) and a smooth positive scalar field β. If $z \in M$, we set $z = (x,t)$ with $x \in M_0$ and $t \in \mathbb{R}$. Then the following Lorentzian metric can be defined on M:

for all $\zeta = (\xi, \tau) \in T_z M = T_x M \times \mathbb{R}$ we set

$$\langle \zeta, \zeta \rangle_L = \langle \xi, \xi \rangle_R - \beta(x) \tau^2 \qquad (1.2)$$

The metric (1.2) is clearly Lorentzian and M, equipped with (1.2) is called *(standard) static space-time*.

Clearly the Minkowski space-time is a static space-time with $M_0 = \mathbb{R}^3$, $\langle \cdot, \cdot \rangle_R$ the Euclidean metric in $\mathbb{R}^3$ and $\beta = 1$.

The coordinates $x \in M_0$, $t \in \mathbb{R}$ of an event z in M are interpreted as "position" and "time" as measured by an observer "far away" from the source of the gravitational field.

A physically relevant example of static Lorentz metric is the Schwarzchild metric which describes the gravitational field in the empty space produced by a nonrotating spherically symmetric body (see section 3).

Now let $\delta : M_0 \longrightarrow TM_0$ be a smooth vector field on M_0 and consider the following Lorentz metric on $M = M_0 \times \mathbb{R}$. For all $\zeta = (\xi, \tau) \in T_z M = T_x M_0 \times \mathbb{R}$ we set

$$\langle \zeta, \zeta \rangle_L = \langle \xi, \xi \rangle_R + 2 \langle \delta(x), \xi \rangle_R \tau - \beta(x) \tau^2. \qquad (1.3)$$

The Lorentz metric (1.3) is called *stationary* and M, equipped with (1.3), is called *stationary space-time*.

A physically relevant example of stationary metric is the Kerr metric which describes the gravitational field in the empty space produced by a rotating spherically symmetric body (see e.g. [28]).

1.3 Space-time of split-type.

Let $(M_0, \langle \cdot, \cdot \rangle_R)$ be a smooth, connected Riemannian manifold. Consider the "time dependent" Lorentz metric $\langle \cdot, \cdot \rangle_L$ on $M = M_0 \times \mathbb{R}$ defined as follows:

$$\langle\zeta,\zeta\rangle_L = \langle A(x,t)\xi,\xi\rangle_R - \beta(x,t)\tau^2 \qquad (1.4)$$

where $z = (x,t) \in M$, $\zeta = (\xi,\tau) \in T_zM = T_xM_0 \times \mathbb{R}$, $A(x,t)$ is a positive linear operator defined on T_xM_0 and β is a smooth, positive scalar field on M.

M, equipped with the Lorentz metric (1.4), is called space-time of *split-type*.

Observe that, when both A and β do not depend on t, a space-time of split-type is static (i.e. of type (1.2)).

The class of the Lorentz manifolds of split-type is quite large. In fact a celebrated result of Geroch (see [19]) shows that any *time-oriented* and *globally hyperbolic* Lorentzian manifold is of split-type. For the definition of time-orientability and global hyperbolicity we refer to [2,28,37].

We recall that the global hyperbolicity is a notion introduced by Leray for the well posedness of the Cauchy problem.

Finally we observe that also the second region of the Reissner-Nördstrom solution of the Einstein equations (see [28]) is a space-time of type (1.4).

2. GEODESICS FOR R-COMPLETE STATIC SPACE-TIMES

Let $M = M_0 \times \mathbb{R}$ be a static space-time (see 1.2). In this section we shall state some existence and multiplicity results for the geodesics under the assumption that M is R-complete. We give the following definition

<u>Definition</u> 2.1 *Let* $M = M_0 \times \mathbb{R}$ *be a static space-time equipped with the metric (1.2). M is called R-complete if* M_0 *is complete as metric space equipped with the distance deduced from the Riemannian metric* $\langle\cdot,\cdot\rangle_R$ *(this is equivalent, by the Hopf-Rinow theorem, to require that any Riemannian geodesic on* M_0 *can be extended for all the values of its parameter).*

As far as the existence and multiplicity of timelike geodesics is concerned, the following theorem can be proved (see [6]).

<u>Theorem 2.2</u> Assume that $M = M_0 \times \mathbb{R}$ is a R-complete static space-time (see Def.2.1). Assume moreover that β (see (1.2) is bounded from above. Let $z_0 = (x_0, t_0)$ and $z_1 = (x_1, t_1)$ be two events in M. Then there exists a timelike geodesic $z = z(s)$ joining z_0 and z_1 if and only if the following assumption holds:

$$\begin{cases} \text{there exists } x \in C^1([0,1], M_0), \ x(0) = x_0, \ x(1) = x_1 \text{ such that} \\ \int_0^1 \langle \dot{x}(s), \dot{x}(s) \rangle_R \, ds \cdot \int_0^1 (1/\beta(x(s))) \, ds < (t_0 - t_1)^2 \end{cases} \quad (2.1)$$

If we assume moreover that M_0 is not contractible in itself, then

$$N(z_0, z_1) \longrightarrow \infty \quad \text{as } |t_0 - t_1| \longrightarrow \infty,$$

where $N(z_0, z_1)$ denotes the number of the timelike geodesics joining z_0 and z_1.

<u>Remark 2.3</u> Clearly for any fixed x_0 and $x_1 \in M_0$, assumption (2.1) is satisfied if we take $t_0 - t_1$ large enough with respect to the Riemann distance $d(x_0, x_1)$ between x_0, x_1. Vice versa (2.1) is not satisfied if $t_0 - t_1$ is small with respect $d(x_0, x_1)$. In particular theorem 2.2 says that two simultaneous events (i.e. with $t_0 = t_1$) cannot be joined by a timelike geodesic.

<u>Remark 2.4</u> The results of theorem 2.2 do not hold, in general, without the assumption that the positive scalar field β is bounded. In fact consider the Anti-de Sitter space-time, i.e. the manifold $M =]-\pi/2, \pi/2[\times \mathbb{R}$ equipped with the static Lorentz metric $ds^2 = \beta(x)(dx^2 - dt^2)$ where dx^2 is the Euclidean metric on $M_0 =]-\pi/2, \pi/2[$ and $\beta(x) = (1/\cos x)^2$. M_0, equipped with the Riemannian metric $(1/\cos x)^2 \cdot dx^2$, is complete then the Anti-de Sitter space-time is R-complete. It can be shown (see [37, sec 1] or [2, sec 5.1]) that there exist events in M which satisfy (2.1) and cannot be joined by any geodesic.

<u>Remark 2.5</u> Theorem 2.1 can be compared with the results of Avez

and Seifert who independently proved the existence of one timelike geodesic joining two suitable related events of a globally Hyperbolic space-time (see [1,40]).

Now consider the problem of the *geodesical connectivity*. We give the following definition:

Definition 2.6 *A space-time M (not necessarily static) is said geodesically connected if for every z_0, z_1 in M there exists a geodesic $z = z(s)$ in M joining z_0 and z_1.*

When we study the geodesical connectivity we need to take into account also the space-like geodesics, since (2.1) is not in general satisfied for two arbitrarily given events z_0, z_1 (see remark 2.3). The following theorem holds (see [6]).

Theorem 2.7 *Any R-complete static space-time M (see def.2.1), whose scalar field β (see (1.2)) is bounded, is geodesically connected (see def.2.6). Moreover, if we assume that M is not contractible in itself, then for any two given events $z_0, z_1 \in M$ there exists a sequence of geodesics $\{z_n(s)\}$ joining z_0 and z_1 such that*

$$\int_0^1 <\dot{z}_n(s), \dot{z}_n(s)>_L ds \longrightarrow +\infty$$

Remark 2.8 The results of theorem 2.7 still hold for stationary space-times (see [5]). The geodesical connectivity of $\mathbb{R}^4$ with respect to a stationary Lorentz metric has been studied in [3].

3. GEODESICS FOR STATIC SPACE-TIMES WITH SINGULAR BOUNDARIES.

In many physically relevant situations the assumption of R-completeness (see def.2.1) of a static Lorentzian manifold is not satisfied due to the presence of a topological boundary ∂M_0 of M_0 on which the Riemannian metric $<\cdot,\cdot>_R$ is singular. Consider for example the solution of the Einstein equation corresponding to the exterior gravitational field produced by a static spherically symmetric body having mass m. This solution, called Schwarzchild metric, can be written (using polar coordinates) in the form (see e.g. [28, 36])

$$ds^2 = (1-(2m/r))^{-1}dr^2 + r^2d\Omega - (1-(2m/r))dt^2 \qquad (3.1)$$

where $d\Omega^2 = d\vartheta^2 + (\sin\vartheta)^2 d\varphi^2$ is the standard metric of the 2-sphere in $\mathbb{R}^3$. The Schwarzschild space-time is the static Lorentz manifold $M = M_0 \times \mathbb{R}$, $M_0 = \{(r,\vartheta,\varphi): r > 2m\}$, equipped with the metric (3.1).

Since (3.1) solves the Einstein equation in the empty space i.e. for $r > r_0$ (r_0 being the radius of the body responsible for the gravitational effects), it is physically meaningful to equip all $M_0 \times \mathbb{R}$ with the the metric (3.1) only if $r_0 \leq 2m$. For a body with the mass of the sun we have $r_0 \leq 3$ Km. Then the Schwarzchild space-time describes situations in which the body has been subject to a gravitational collapse. Notice that the Riemannian metric

$$dx^2 = (1-(2m/r))^{-1}dr^2 + r^2 d\Omega \qquad (3.2)$$

is singular on the boundary $\partial M_0 = \{(r,\vartheta,\varphi): r = 2m \}$. Moreover M_0, equipped with the Riemannian metric (3.2), is not complete. Indeed a radial geodesic $r = r(s)$ for (3.2) which starts at $\rho > 2m$ and "reachs" the boundary ∂M_0, has finite length

$$l(\rho) = \int_{2m}^{\rho} \frac{dr}{(1-2m/r)^{1/2}}.$$

Moreover, since $r = r(s)$ is a geodesic, its affine parameter s is proportional to the arclenght, then $r = r(s)$ ''reaches'' the boundary for a finite value s_0 of the parameter. Then we conclude that the Schwarzchild space-time is not R-complete.

Nevertheless it can be shown that the spatial component $x(s)$ of a timelike geodesic in the Schwarzchild space-time can "reach" the boundary ∂M_0 only if the time-coordinate $t(s)$ goes to infinity. A proof of this fact is contained in a following remark (see remark 3.3).

Having in mind, as model, the Schwarzchild space-time we are led to introduce the following definition (see [7])

<u>Definition 3.1</u> *Let $M = M_0 \times \mathbb{R}$ be an open, connected subset of a smooth manifold. Let $\partial M = \partial M_0 \times \mathbb{R}$ denote the (topological) boundary of M and equip M with a static Lorentz metric (1.2). M is called "Universe (with a black hole)" if the following assumptions are satisfied:*

$$\beta(x) \longrightarrow 0 \quad \text{as } x \longrightarrow x_0 \in \partial M_0 \tag{3.3}$$

— *For any $\delta > 0$ the set $\{x \in M_0 : \beta(x) \geq \delta\}$ is complete (with respect to the Riemann metric on M_0).* (3.4)

— *For any timelike geodesic $z(s)=(x(s),t(s))$, $s \in [0, s_0[$, in M such that $\liminf \beta(x(s)))=0$ as $s \longrightarrow s_0$, we have*

$$\limsup_{s \to s_0} |t(s)| = +\infty \tag{3.5}$$

<u>Remark 3.2</u> Observe that condition (3.5) can be physically interpreted as follows: if a material particle reaches the topological boundary ∂M_0, within a finite proper time s_0, an observer far from the boundary does not see this event in finite time t. Roughly speaking (3.5) means that an observer (far away from the gravitational effects) does not see anything "to go in" or "to go out" from M_0. This fact justifies the name of "Universe with a black hole" for the structure introduced in definition 3.1.

<u>Remark 3.3</u> If M_0 is complete with respect to the Riemannian metric $\langle \cdot, \cdot \rangle_R / \beta(x)$, then (3.5) holds. In fact let $(x(s), t(s))$ be a timelike geodesic, so for all s we have

$$\langle \dot{x}(s), \dot{x}(s) \rangle < \beta(x(s)) \dot{t}^2(s).$$

From this inequality and the monotonicity of t(s) (which can be directly deduced from the geodesic equation) we easily have

$$|t(s)| > \int_0^S (<\dot{x}(s),\dot{x}(s)>/\beta(x(s)))^{1/2}ds - |t(0)|$$

The above inequality and the completeness of the metric $<\cdot,\cdot>_R/\beta(x)$ imply (3.5).

Observe that the Schwarzchild space-time satisfies (3.5) since $M_0 = \{x \in \mathbb{R}^3 : |x| > 2m\}$, with the metric $dx^2/(1-(2m/r))$ (dx^2 being defined in (3.2)), is clearly complete.

For Universes with black holes we can prove the same existence and multiplicity results contained in theorem 2.1. More precisely the following theorem holds (see [7]):

Theorem 3.4 Let M be an Universe (see def. 3.1) whose scalar field β (see (1.2)) is bounded. Then if assumption (2.1) is satisfied the same conclusions of theorem 2.2 hold.

For the study of the geodesical connectivity the condition that M is an Universe is not appropriated. Indeed consider the Lorentz manifold $M = M_0 \times \mathbb{R}$, $M_0 = \{x = (x_1,x_2,x_3) \in \mathbb{R}^3 : x_1^2 + x_2^2 + x_3^2 > 1\}$ with the metric

$$ds^2 = dx^2 - \beta(x)dt^2 \qquad (3.6)$$

where dx^2 is the euclidean metric on $\mathbb{R}^3$ and $\beta(x)$ is a smooth, bounded, positive scalar field on M_0 such that $\beta(x) = (|x|-1)^2$ if $|x| \leq 2$. It is not difficult to see that M_0, equipped with the metric $dx^2/\beta(x)$, is complete. So, by using remark 3.3, we deduce that M with the metric (3.6) is a universe with a black hole. Nevertheless it is not geodesically connected since it is easy to see that the only geodesic for (3.6) joining the two simultaneous events $(x,0)$, $(-x,0)$, $x \in M_0$, is a straight line segment which is not contained in M.

For this reason we introduce a condition of convexity, with respect to the Lorentz metric, for the boundary ∂M of M. In order to introduce this definition we need some notations.

Let $(M, <,>_L)$ be a space-time (not necessarily static). Then :

- grad $\phi(z)$ denotes the gradient of the scalar field ϕ on M with respect to the Lorentz structure, i.e. it is the unique vector field G on T_zM such that $\langle G,\zeta\rangle_L = d\phi(z)\zeta$ for all $\zeta \in T_zM$.
- $H^\phi_L(z)[\zeta,\zeta]$ denotes the (Lorentz) Hessian of ϕ at $z \in M$ in the direction $\zeta \in T_zM$, i.e.

$$H^\phi_L(z)[\zeta,\zeta] = \frac{d^2}{ds^2}(\phi(z(s)))_{s=0},$$

where $z = z(s)$ is the geodesic (for the Lorentz metric) such that $z(0) = z$ and $\dot z(0) = \zeta$.

We are now ready to give the following definition (see [7])

Definition 3.6 Let $M = M_0 \times \mathbb{R}$ be an open, connected subset of a smooth manifold. Assume that on M a static space-time metric is given. The (topological) boundary $\partial M = \partial M_0 \times \mathbb{R}$ is said to be convex if there exists a smooth positive map ϕ on M having the following properties:

i) $\phi(x,t) = \phi(x)$ for all $(x,t) \in M$.

ii) $\phi(x,t) \longrightarrow 0$ as $(x,t) \longrightarrow z \in \partial M$.

iii) For any $\delta > 0$ the set $\{x \in M_0 : \beta(x) \geq \delta\}$ is complete (with respect to the Riemann metric on M_0).

iv) There exists a neighborhood N of ∂M and positive constants n,m,p such that for all $z \in (N \cap M)\setminus\partial M$ we have

$n \geq \langle\text{grad } \phi(z),\text{grad } \phi(z)\rangle_L \geq m$

and

$H^\phi_L(z)[\zeta,\zeta] \leq p|\langle\zeta,\zeta\rangle_L| \phi(z)$ for all $\zeta \in T_zM$

Remark 3.7 It is possible to prove, (see [7]), that the Schwarzschild space-time satisfies iv) of definition 3.6 with the function $\phi(r,\vartheta,\varphi,t) = (1-(2m/r))^{1/2}$. Then the Schwarzschild space-time has convex boundary.

Remark 3.8 In order to interpret definition 3.6 notice that ϕ tends to zero on ∂M. Then iv) implies that for all $z_0 \in \partial M$ we have

$$\limsup_{z \to z_0} H_L^\phi(z)[\zeta,\zeta] \leq 0, \text{ for all } \zeta \in T_z M \text{ with } |<\zeta,\zeta>_L| \leq 1 \quad (3.7)$$

Now let ϕ be a smooth, real map on the Euclidean space $\mathbb{R}^n$ whose zeros forms an n-1 dimensional manifold. Consider the subset A defined by

$$A = \left\{ z \in \mathbb{R}^n \mid \phi(z) \geq 0 \right\}$$

It is easy to see that, if A is convex (with the usual meaning), then the (Euclidean) Hessian $H^\phi(z)$ of ϕ is negative semidefinite for all $z \in \partial A$ (compare with (3.7)).

Unlike the case described above, in the "setting" of definition 3.6 the static lorentz metric is not defined on ∂M (and this indeed happens for the Schwarzchild space-time (see (3.1)). Thus we cannot consider the Lorentzian Hessian on ∂M but only "near" ∂M, namely on $(N \cap M)\backslash\partial M$ (see iv in def.3.6).

. The convexity assumption in definition 3.6 permits us to get the results of theorem 2.7 for static space-times which are not R-complete. More precisely the following theorem holds (see [7]).

Theorem 3.9 Let M be a static space-time having convex boundary (def.3.6) and whose scalar field β (see (1.2)) is bounded. Then the same conclusions of theorem 2.7 hold.

Remark 3.10 The statements of theorem 3.9 still hold, with suitable modifications, for stationary space-times (see [21]).

4. OUTLINE OF THE PROOF OF THEOREMS

The proofs of the theorems stated in the preceding sections utilize a variational principle which reduces the study of geodesics for static space-time to the study of a suitable functional which is bounded from below (see (4.4)). First we introduce some notation.

Let $M = M_0 \times \mathbb{R}$ be a static space-time with the metric (1.2). Let $z_0 = (x_0, t_0)$, $z_1 = (x_1, t_1)$ two events in M. We shall consider path spaces on M_0 and M defined in the following way:

$$\Omega^1 = \Omega^1(M_0, x_0, x_1) = \{x : [0,1] \longrightarrow M_0 : x \text{ absolutely continuous}$$

and such that $\int_0^1 <\dot{x}, \dot{x}>_R ds < \infty$, $x(0) = x_0, x(1) = x_1\}$. (4.1)

$$\Gamma^1 = \Omega^1 \times H^1(t_0, t_1). \quad (4.2)$$

$$H^1(t_0, t_1) = \{t \in H^1([0,1], \mathbb{R}) : t(0) = t_0, t(1) = t_1\} \quad (4.3)$$

Here $H^1([0,1], \mathbb{R})$ denotes the Sobolev space of the absolutely continuous curves whose derivative is square summable.

Both Ω^1 and Γ^1 are Riemann manifolds, modeled on Sobolev spaces, of curves in $\mathbb{R}^3$ and in $\mathbb{R}^4$ respectively (see for instance [29,33]). For every $x \in \Omega^1$ the tangent space to Ω^1 is

$T_x\Omega^1 = \{\xi = \xi(s), s \in [0,1]: \xi \text{ absolutely continuous vector field}$

along x such that $\int_0^1 <D_s\xi, D_s\xi>ds < \infty$ and $\xi(0) = \xi(1) = 0\}$.

For every $z = (x,t) \in \Gamma^1$ the tangent space to Γ^1 is

$$T_z\Gamma^1 = T_x\Omega^1 \times H_0^1, \quad H_0^1 = \{t \in H^1([0,1], \mathbb{R}) \text{ such that } t(0) = t(1) = 0\}$$

The geodesics joining z_0 and z_1 are the critical points of the functional f on Γ^1 defined by

$$f(z) = \int_0^1 <\dot{z}(s), \dot{z}(s)>_L ds = \int_0^1 [<\dot{x}(s), \dot{x}(s)>_R - \beta(x(s))\dot{t}(s)^2] ds \quad (4.4)$$

The study of the critical points of (4.4) can be reduced to the study of a suitable functional J which is bounded from below and which is defined only on the "spatial" manifold Ω^1. In fact the following variational principle holds (see [6])

<u>Theorem 4.1</u> Let $M = M_0 \times \mathbb{R}$ be a static space-time with the metric (1.2). Let $z_0 = (x_0, t_0)$ and $z_1 = (x_1, t_1)$ be two events in M. A curve $z(s) = (x(s), t(s)) \in \Gamma^1$ (see (4.2)) is a critical points of f (see (4.4) if and only if

$$t(s) = \Phi(x)(s) := t_0 + (t_0 - t_1)\left(\int_0^1 (1/\beta(x))ds\right)^{-1}\int_0^s (1/\beta(x(r)))dr \quad (4.5)$$

and $x = x(s)$ *is a critical point of the functional J on* Ω^1 *(see (4.1)) defined by*

$$J(x)=f(x,\phi(x))=\int_0^1 (\langle \dot{x}(s),\dot{x}(s)\rangle_R \, ds - (t_0-t_1)^2 \left(\int_0^1 (1/\beta(x))ds \right)^{-1} \quad (4.6)$$

Moreover if $z(s) = (x(s),t(s))$ *is a geodesic we have*

$$f(z) = J(x) \quad (4.7)$$

Proof. For any fixed x in Ω^1 the functional $f(x,t)$ in (4.4) is concave in $t \in H^1(t_0,t_1)$. In fact a direct calculation shows that for all $t \in H^1(t_0,t_1)$ and $\tau \in H_0^1$ we have

$$f''_{tt}(x,t)[\tau,\tau] = -2\int_0^1 \beta(x(s))\dot{\tau}(s)^2 ds$$

where $f''_{tt}(x,t)$ denotes the second frechet differential of f with respect to t.

Then, by the implicit function theorem, the set $N = \{(x,t): f'_t(x,t)=0\}$ ($f'_t(x,t)$ denotes the partial differential with respect to t) is the graph of a smooth function $\Phi : \Omega^1 \longrightarrow H^1(t_0,t_1)$. Now consider the functional on Ω^1 defined by

$$J(x) = f(x,\Phi(x)) \qquad x \in \Omega^1 \quad (4.8)$$

A direct use of the chain rule shows that (x,t) is a critical point of f if and only if x is a critical point of J and $t=\Phi(x)$.

Now it is to verify that Φ and J are defined respectively by (4.5) and (4.6). Moreover (4.7) follows from definition (4.8).

Remark. It is easy to see that the functional (4.6) is bounded from below if the scalar field β is bounded from above.

Sketch of the proof of theorem 2.2

Let $z(s) = (x(s),t(s)) \in \Gamma^1$ be a timelike geodesic. Then, by (4.7), we have $J(x) < 0$, and this inequality implies (2.1). Vice versa now assume that (2.1) holds. Since $(M_0,\langle\cdot,\cdot\rangle_R)$ is complete, the path manifold Ω^1 is complete too; then since the functional J is bounded from below it is not difficult to show that J has a minimum x in Ω^1. Clearly, by theorem 4.1, z = (x,t), with t defined by (4.5), is a geodesic joining z_0, z_1. Moreover, since x is a minimum for J in Ω^1 and assumption (2.1) holds, we clearly get J(x) < 0. This implies, by (4.6), that z is timelike. Then the first part of theorem 2.1 is proved.

Now assume that M_0 is not contractible in itself. In this case the path manifold Ω^1 has a "rich" topology. Namely (see [15]) there exists a sequence $\{K_n\}$ of compact subsets of Ω^1 such that

$$\text{cat } (K_n, \Omega^1) \longrightarrow \infty \qquad \text{as } n \longrightarrow \infty \qquad (4.9)$$

where cat (K_n, Ω^1) denotes the Lusternik-Schnirelmann category of K_n with respect Ω^1 (see e.g. [39,41]), i.e. the minimal number of closed, contractible (in Ω^1) subsets of Ω^1 covering K_n.

Now consider the sublevel $J^{-1} = \{x \in \Omega^1 : J(x) \le -1\}$. By the definition (4.6) of J and by (4.9) it is easy to deduce that

$$\text{cat } (J^{-1}, \Omega^1) \longrightarrow \infty \qquad \text{as } |t_0 - t_1| \longrightarrow \infty \qquad (4.10)$$

Moreover it can be shown that the functional J satisfies the Palais-Smale compactness condition. Then, by (4.10) and by well known arguments of the Lusternik-Schnirelmann theory (see e.g. [39,41]), we get the multiplicity result for the critical points of the functional J. These critical points have energy ≤ -1, then they, by (4.7) they give rise timelike geodesics.

Sketch of the proof of theorem 3.4

Since M_0 with the Riemannian metric $\langle \cdot, \cdot \rangle_R$ is not complete, the path manifold Ω^1 is not complete. To overcome this lack of completeness we introduce, for $\varepsilon > 0$, the "penalized" functional

$$J_\varepsilon(x) = J(x) + \varepsilon \int_0^1 (1/\beta(x))^2 ds, \qquad x \in \Omega^1 \qquad (4.11)$$

where J(x) has been defined in (4.6).

Now, since $\beta(x)$ goes to zero on the boundary ∂M (see (3.3)), it can be shown that the curves x in the sublevel $J_\varepsilon^a = \{x \in \Omega^1 : J_\varepsilon(x) \le a\}$ ($a \in R$) remain (uniformly with $x \in J_\varepsilon^a$) "far away" from the boundary ∂M_0 (i.e. $\beta(x(s)) \ge \delta$, with $\delta > 0$ independent on x in

J_ε^a and $s \in [0,1]$). Then, by (3.4), the sublevels J_ε^a are complete and it can be shown that, for any $\varepsilon > 0$, J_ε has a minimum $x_\varepsilon \in \Omega^1$. If $\bar{x}$ is the curve in (2.1), clearly we have for all ε small

$$J_\varepsilon(x_\varepsilon) \le J_\varepsilon(\bar{x}) \le J(\bar{x}) < 0 \qquad (4.12)$$

(4.12) implies that $\{x_\varepsilon : 0 < \varepsilon < 1\}$ is bounded in the H^1 norm. Moreover, by (4.12) and exploiting (3.5), it is possible to prove that $\beta(x_\varepsilon(s)) \ge \delta$, with $\delta > 0$ independent on ε and s; then (passing eventually to a subsequence), by (3.4), x_ε converges weakly in H^1 as $\varepsilon \longrightarrow 0$ to a curve x s.t. for all s $\beta(x(s)) \ge \delta$. Now passing to the limit $\varepsilon \longrightarrow 0$ in the equation $dJ_\varepsilon(x_\varepsilon) = 0$ (dJ_ε denotes the differential of J_ε) it can be shown that x solves the equation $dJ(x) = 0$. Then, by lemma 4.1, $z = (x,t)$, with t defined by (4.5), is a geodesic joining z_0 and z_1. Finally (4.12) implies that z is timelike.

In order to get the multiplicity result when M_0 is not contractible in itself, we need, as in the proof of theorem 2.1, to exploit the "richness" of the topology of the path manifold Ω^1. We need also to take a different penalization in (4.11) in order to preserve the multiplicity of the critical points of J_ε when we pass to the limit $\varepsilon \longrightarrow 0$.

Sketch of the proof of theorem 2.7

Let z_0, z_1 be two points in M. The existence of a geodesic joining z_0, z_1 can be proved by a minimum argument for the functional J.

Now we assume that M_0 is not contractible in itself and prove the multiplicity result. First observe that J satisfies the Palais-Smale condition. Then for any real number c it can be shown that

$$n \equiv \text{cat } (J^c, \Omega^1) < \infty$$

Then, by the properties of the L.S. category, we have:

$$c(n) \equiv \inf\{ \sup (J(A)): A \subseteq \Omega^1, \text{ cat } (A,\Omega^1) \ge n+1 \} \ge c$$

Moreover, by standard arguments of the Lusternik-Schnirelmann critical point theory, $c(n)$ is a critical value of J. Finally

, since c is arbitrary, the conclusion follows.

Sketch of the proof of theorem 3.9

As in the of theorem 3.4, to overcome the lack of completeness of Ω^1, we consider the penalized functional

$$J_\varepsilon(x) = J(x) + \varepsilon \int_0^1 (1/\Phi(x))^2 ds, \qquad x \in \Omega^1$$

where $J(x)$ has been defined in (4.6) and $\Phi(x)$ is the scalar field on M defining the boundary ∂M (see Def.3.6). Arguing exactly as in the proof of theorem 3.4, it can be proved that for any $\varepsilon > 0$, J_ε has a minimum $x_\varepsilon \in \Omega^1$.
Clearly

$$\{J_\varepsilon(x_\varepsilon): 0 < \varepsilon \leq 1\} \quad \text{is bounded.} \tag{4.13}$$

Now, exactly as in the proof of theorem 3.4, the main step in the research of a critical point of J consists in proving that the curves x_ε are uniformly (with ε) far away from the boundary ∂M, namely that

$$\Phi(x_\varepsilon(s)) \geq \delta, \text{ with } \delta > 0 \quad \text{independent on } \varepsilon \text{ and } s. \tag{4.14}$$

(4.14) is a consequence of (4.13) and of the convexity assumption of the boundary (see def.3.6). In fact, arguing by contradiction, assume that, for a suitable sequence $\{s_\varepsilon\}$ in $[0,1]$, we have

$$u_\varepsilon(s_\varepsilon) \longrightarrow 0 \qquad u_\varepsilon(s) \equiv \Phi(x_\varepsilon(s)) \tag{4.15}$$

Using the convexity of ∂M, it possible to prove that

$$\ddot{u}_\varepsilon(s) \leq \text{const.} u_\varepsilon(s)$$

Then, by the Gronwall lemma and (4.15), we deduce that u_ε converges uniformly to 0 and this contradicts the boundary conditions $u_\varepsilon(0) = \Phi(x_0) > 0$, $u_\varepsilon(1) = \Phi(x_1) > 0$.
The multiplicity result can be obtained by using in a suitable way the Lusternik-Schnirelmann critical point theory.

5. GEODESICS FOR TIME-DEPENDENT SPACE-TIMES OF SPLIT-TYPE.

In the previous sections we have examined the existence of

geodesics in static space-times. Here we consider "time-dependent" space-times of split-type (see 1.3). Namely we consider a connected manifold

$$M = M_0 \times \mathbb{R}$$

equipped with the Lorentz metric

$$\langle \zeta, \zeta \rangle_L = \langle A(x,t)\xi, \xi \rangle_R - \beta(x,t)\tau^2 \tag{5.1}$$

for all $z = (x,t) \in M$, $\zeta = (\xi,\tau) \in T_zM = T_xM \times \mathbb{R}$, where $\langle\,,\,\rangle_R$ is a Riemannian metric on M_0, $A(x,t)$ is a positive linear operator defined on T_xM_0 and β is a smooth, positive scalar field on M.

The following theorem holds (see [12])

<u>Theorem</u> 5.1 Let $M_0 \times \mathbb{R}$ be a lorentzian manifold of split-type. Assume that:

(i) $(M_0, \langle\,,\,\rangle_R)$ is a complete Riemannian manifold.

(ii) $\nu \langle \xi, \xi \rangle_R \leq \langle A(x,t)\xi, \xi \rangle_R \leq M \langle \xi, \xi \rangle_R$; $\nu \leq \beta(x,t) \leq M$

where ν and M are positive constants.

(iii) $\frac{\partial}{\partial t} A(x,t)$ and $\frac{\partial}{\partial t} \beta(x,t)$ are bounded.

(iv) $\limsup_{t \to +\infty} \langle \frac{\partial}{\partial t} A(x,t)\xi, \xi \rangle_R \leq 0$

$\liminf_{t \to -\infty} \langle \frac{\partial}{\partial t} A(x,t)\xi, \xi \rangle_R \geq 0$

uniformly in $x \in M_0$ and $\xi \in T_xM_0$.

Then $(M, \langle\,,\,\rangle_L)$ is geodesically connected. Moreover, if we assume that M is not contractible in itself, then for any two given events $z_0, z_1 \in M$ there exists a sequence of geodesics $\{z_n(s)\}$ joining z_0 and z_1 such that

$$\int_0^1 \langle \dot{z}_n(s), \dot{z}_n(s) \rangle_L \, ds \longrightarrow +\infty$$

For a proof of this theorem, under slight different assumptions, we refer to [31]. Here we shall only observe that the main difficulty in the proof of theorem 5.1 is due to the fact that we have not been able to prove the Palais-Smale condition (actually we suspect that the energy functional related to a "time dependent" space-time of split-type does not satisfy this assumption in general).

Remark 5.2 A space-time $M = M_0 \times \mathbb{R}$ of split-type which does not satisfy the assumptions of theorem 5.1 is not in general geodesically connected. Consider for example $\mathbb{R}^2$ equipped with the Lorentzian metric
$$ds^2 = (1 + t^2)dx^2 - (1/(1 + t^2))dt^2$$
Notice that after the change of variable $t = tg(\tau)$ we get (with x and τ interchanged) the the Anti-de Sitter space-time which is not geodesically connected (see Remark 2.4).

Remark 5.3 Under the assumptions of theorem 5.1 the existence of a timelike geodesic joining $z_0 = (x_0, t_0)$ and $z_1 = (x_1, t_1)$ can be proved provided $|t_0 - t_1|$ is sufficiently large.

Remark 5.4 Let us observe that it is also interesting to study "time-dependent" space-times of split-type under different assumptions than those occurring in theorem 5.1. For example the the "second region" of the Reissner-Nördstrom space-time (see [28]) does not satisfy the assumptions in theorem 5.1. The geodesical connectivity of Lorentz manifolds of this type has been studied in [20,22].

6. MORSE THEORY FOR LORENTZIAN MANIFOLD.

In this section we shall develop a Morse theory for geodesics which can be also spacelike) in Lorentzian manifolds of plit-type satisfying assumptions which are slightly more 1estrictive than those in theorem 5.1.

To this end it will be useful to recall some basic facts about the Morse theory for geodesics on Riemannian manifolds.

Morse theory for Riemannian manifolds.

Let $(M_0, <\cdot,\cdot>_R)$ be a smooth, complete Riemannian manifold. Given two points x_0, $x_1 \in M_0$ the geodesics joining x_0 and x_1 are the critical points of the energy integral

$$f(x) = \int_0^1 <\dot{x}(s),\dot{x}(s)>_R ds,$$

defined on the path manifold Ω^1 (see 4.1).

Let x be a geodesic joining x_0, x_1, then the Morse index m(x) is the maximal dimension of a subspace of $T_x\Omega^1$ (the tangent space at x to Ω^1) on which the hessian $f''(x)$ is negative definite. We recall that the hessian $f''(x): T_x\Omega^1 \times T_x\Omega^1 \longrightarrow \mathbb{R}$ of f at the geodesic x is given by

$$f''(x)[\xi,\xi'] = \int_0^1 [<D_s\xi, D_s\xi'>_R - <R(\dot{x},\xi)\dot{x},\xi'>_R] ds \qquad (6.1)$$

where R is the tensor curvature of $<\cdot,\cdot>_R$ (see for instance [29]). Notice that m(x) is finite because $f''(x)$ is a compact perturbation of the positive definite bilinear form

$$a(x)[\xi,\xi'] = \int_0^1 (<D_s\xi,D_s\xi'>_R ds.$$

For every geodesic x, there is another index $\mu(x)$, which we call "geometric index". We need some notation. For every $s \in]0,1]$, we consider the functional

$$f_s(y) = \int_0^s <\dot{y}(\sigma),\dot{y}(\sigma)>_R d\sigma,$$

defined on the path manifold $\Omega_s^1 = \Omega^1(x_0, x(s), M_0)$ of the curves joining x_0 and $x(s)$ and parameterized in $[0,s]$. Moreover, let x_s be the restriction of x to $[0,s]$. Then x_s is a critical point of f_s.

Definition 6.1 Let x be a geodesic joining x_0, x_1. A point x(s), s $\in]0,1]$, is said conjugate to x_0 along x, if $f''(x_s)$ is degenerate, i.e. there exists $\xi \in T_{x_s}\Omega^1$, $\xi \neq 0$, such that

$$f''(x_s)[\xi,\xi'] = 0, \qquad \text{for all } \xi' \in T_{x_s}\Omega^1$$

ξ is said Jacobi tangent field along x_s.

The set of the Jacobi tangent fields along x_s is the kernel of $f''(x_s)$. The multiplicity of the conjugate point $x(s)$ is the dimension of Ker $f''(x_s)$. It is a finite number. Indeed, from (6.1), a vector field ξ belongs to Ker $f''(x_s)$ if and only if it solves the following system of linear ordinary differential equations

$$D_s^2\xi + R(\dot{x},\xi)\dot{x} = 0 \qquad \xi(0) = \xi(s) = 0$$

Then Ker $f''(x_s)$ has dimension at most $n = \dim M_0$ (actually at most $n-1$, see [35]).

Definition 6.2 *The geometric index $\mu(x)$ of a geodesic x in M_0 and parameterized in $[0,1]$ is the number of points $x(s)$, $s \in \,]0,1[$, conjugate to x_0 and counted with their multiplicity.*

The Morse index theorem is one of the most important in global Riemannian geometry (for the proof see [29, 35]).

Theorem 6.3 *For every geodesic x we have $m(x) = \mu(x)$.*

As an immediate consequence we have that a geodesic $x: [0,1] \longrightarrow M_0$ can contain only finitely many points which are conjugate to $x(0)$ along x. Then, if s is small enough, for x_s (x_s being the restriction of x to $[0,s]$) we have $m(x_s) = \mu(x_s) = 0$ and Ker $f''(x_s) = \{0\}$. So $f''(x_s)$ is positive definite and consequently the geodesic arc x_s is a local minimum for f_s.

Let x_0, x_1 be two nonconjugate points of M_0, i.e. they are nonconjugate along every geodesic joining them. We incidentally observe that, by Sard theorem, almost all couples of points of M_0 are nonconjugate. Indeed a point q is conjugate to a point p along a geodesic if and only if q is a critical value of the exponential map $\exp_p$ (see for instance [35, Theorem 18.1]). For every $c \in \mathbb{R}$, we consider the sublevels $f^c = \{x \in \Omega^1 : f(x) \leq c\}$ of the energy functional. Moreover, if $H_k(f^c, \mathbb{R})$ is the k-th singular homology group of f^c (with real coefficients), we consider the Betti numbers $b(k,c) = \dim H_k(f^c, \mathbb{R})$. Since for every $c \in \mathbb{R}$ f^c is homotopically equivalent to a compact manifold (see for instance [35, theorem 16.2]), the numbers b(k,c) are finite and definitively null. Let us now recall the Morse relations for the energy functional of two nonconjugate points (see for instance [13]).

Theorem 6.4 *Let c be a noncritical value of f (i.e. there are not critical points at level c). Moreover we set* $Z_c = \{x \in f^c: f'(x) = 0\}$. *We have:*

$$\sum_{x \in Z_c} t^{\mu(x)} = \sum_{k=0}^{\infty} b(k,c) t^k + (1+t)Q(t) \qquad (6.2)$$

where Q(t) is a polynomial with positive integer coefficients.

Remark 6.5 Notice that the sum on the left of (6.2) is finite. Indeed, since x_0, x_1 are nonconjugate, f is a Morse function (i.e. for every critical point x of f, f"(x) is nondegenerate). Then every critical point is isolated. Moreover f satisfies the Palais-Smale compactness condition, then f has a finite number of critical points on every sublevel.

We finally recall that (6.3) is an immediate consequence of the Morse relations i.e.

$$\sum_{x \in Z_c} t^{m(x)} = \sum_{k=0}^{\infty} b(k,c) t^k + (1+t)Q(t)$$

and of theorem 6.3.

Now we are ready to state the results about the Lorentz manifolds.

Statement of the result.

The definitions of conjugate points, Jacobi tangent fields, Morse index and geometrical index can be transferred to Lorentz metric; of course $\langle \cdot, \cdot \rangle_R$ needs to be replaced by $\langle \cdot, \cdot \rangle_L$. Thus it make sense to ask ourselves if the analogous of theorem 6.4 holds in some Lorentz manifolds.

Here we consider "time dependent" space-times of split type as in the preceding section (see 5.1).

For every geodesic $z = (x,t)$ and $\zeta = (\xi,\tau)$, $\zeta' = (\xi',\tau') \in T_z\Gamma^1$, the Hessian $f''(z): T_z\Gamma^1 \times T_z\Gamma^1 \longrightarrow \mathbb{R}$ is given by

$$f''(z)[\zeta,\zeta'] = \int_0^1 [<D_s\zeta, D_s\zeta'>_L - <R_L(\dot x,\zeta)\dot x,\zeta'>_L]ds,$$

Here R_L is the curvature tensor with respect to the Lorentz metric $<\cdot,\cdot>_L$.

Unfortunately the Morse index is $+\infty$ because $f''(z)$ is negative definite on the "time" direction $(0,\tau)$, $\tau \in H^1([0,1],\mathbb{R})$ with $\tau(0) = \tau(1) = 0$ (this can be easily checked by taking the sequence $(0,\sin(kts))$, k positive integer sufficiently large).

However it can be shown that the geometrical index $\mu(z)$ is finite, as in the Riemannian case, and this index can be exploited to construct a Morse theory.

The following theorem holds (see [10]):

<u>Theorem 6.6</u> *Let $M = M_0 \times \mathbb{R}$ be a space-time of split-type satisfying all the assumptions of theorem 5.1. Assume moreover that*

$$\frac{\partial^2}{\partial t^2} A(x,t), \frac{\partial^2}{\partial t^2} \beta(x,t) \quad \text{are bounded.}$$

Let z_0, z_1 be two nonconjugate points of M and Z be the set of the geodesics joining them. Then

$$\sum_{z \in Z} t^{\mu(z)} = \sum_{k=0}^{\infty} b(k,\Gamma^1)t^k + (1+t)Q(t) \qquad (6.3)$$

where $b(k,\Gamma^1) = \dim H_k(\Gamma^1,\mathbb{R})$ is the k-th Betti number of the loop space Γ^1 and $Q(t)$ is a formal series with positive integer coefficients.

Notice that in the sum (6.3), it is possible that some coefficients are infinite cardinal numbers. In this case, the sum makes sense with the usual algebra of infinite numbers.

<u>Remark 6.7</u> Since the manifold $\{t \in H^1([0,1],\mathbb{R}): t(0) = t_0, t(1) =$

$t_1\}$ is contractible to a point, Γ^1 and Ω^1 are homotopically equivalent. Hence, for every $k \in \mathbb{N}$, $H_k(\Gamma^1,\mathbb{R}) \cong H_k(\Omega^1,\mathbb{R})$. Then, if M_0 is not contractible in itself, for infinitely many $k \in \mathbb{N}$ $b(k,\Gamma^1) = \dim H_k(\Gamma^1,\mathbb{R}) = \dim H_k(\Omega^1,\mathbb{R}) \neq 0$ (see [15]). Therefore in this case we deduce from (6.3) that there are infinitely many geodesics joining any two nonconjugate points in M.

Remark 6.8 Previous results about a Morse theory for geodesics in Lorentz manifolds have been obtained in [42] (see also [2] and its references). In [2,42] only causal (i.e. non space-like) geodesics are considered. Therefore the index of causal geodesics reflects only the topology of the path manifold of the causal curves. Moreover in [42] also a Morse theory of light-like geodesics joining a point with a given time-like curve $z = z(s)$ (i.e. with $\langle \dot z(s), \dot z(s) \rangle_L < 0$ for all s) is presented.

Outline of the proof of theorem 6.6.

The proof theorem 6.6 is very technical and we refer to [10] for it. Here we consider the particular case of static space-times, i.e. we assume that $A(x,t)=$identity and β does not depend on t (see (5.1)). This case has been studied in [11].

Since we consider a static space-time, we can use theorem 4.1. Let $\bar x$ be a critical point of the functional J defined by (4.6).

Then (see theorem 4.1) $\bar z = (\bar x, \bar t)$, with $\bar t = \Phi(\bar x)$ (see (4.5)), is a geodesic joining z_0, z_1.

As in the Riemannian case, for every $s \in]0,1]$ we consider the functional

$$f_s(w) = \frac{1}{2} \int_0^s \langle \dot w, \dot w \rangle_L \, dr,$$

defined on the loop space $\Gamma_s^1 = \Omega_s^1 \times H^1(t_0, t(s))$ of the curves joining z_0 and $\bar z(s)$, and parameterized in $[0,s]$. Then the restriction $\bar z_s$ of $\bar z$ to $[0,s]$ is a critical point of f_s. Now consider the map

$\phi_S: \Omega_S^1 \longrightarrow H_0^1(t_0, t(s))$ defined by

$$\phi_S(y)(r) = t_0 + \frac{\Delta_S}{\int_0^S \frac{1}{\beta(y(\rho))} d\rho} \cdot \int_0^r \frac{1}{\beta(y(\rho))} d\rho,$$

where $\Delta_S = t(s) - t_0$. Then we set

$$J_S(y) = f_S(y, \phi_S(y)) = \frac{1}{2} \int_0^S \langle \dot{y}, \dot{y} \rangle dr - \frac{\Delta_S^2}{2} \frac{1}{\int_0^S \frac{1}{\beta(y(r))} dr}.$$

Now, we generalize the notions of conjugate point and Jacobi tangent field for the critical point $\bar{x}$ of J.

A point $\bar{x}(s)$, $s \in \,]0,1]$, is said conjugate to x_0 along $\bar{x}$ if $J_S''(\bar{x}_S)$ is degenerate, i.e. there exists $\xi \in T_{\bar{x}_S} \Omega_S^1$, $\xi \neq 0$, s.t.

$$J_S''(\bar{x}_S)[\xi, \xi'] = 0, \qquad \forall \xi' \in T_{\bar{x}_S} \Omega_S^1.$$

ξ is said Jacobi tangent field along $\bar{x}_S$ for J. The set of Jacobi tangent fields along $\bar{x}_S$ is the kernel of $J_S''(\bar{x}_S)$ and is denoted by Ker $J_S''(\bar{x}_S)$. The *multiplicity* of the conjugate point $\bar{x}(s)$ is the dimension of Ker $J_S''(\bar{x}_S)$. Finally the *geometrical index* $\mu(\bar{x})$ is the number of conjugate points to x_0 along $\bar{x}$, counted with their multiplicity.

The following "second order" variational principle holds

Theorem 6.9 Let $z = (x, t)$ be a geodesic and $s \in \,]0,1]$, then we have:

i) $\zeta = (\xi, \tau)$ is a Jacobi tangent field (for f) along z_S iff $\tau = \phi_S'(x)\xi$ and ξ is a Jacobi tangent field (for J) along x_S;

ii) $z(s)$ is conjugate to $z(0)$ along z iff $x(s)$ is conjugate to $x(0)$ along x;

iii) $\mu(z) = \mu(x)$.

The proof of the theorem 6.9 can be obtained by a direct calculation.

Since J is a compact perturbation of the Riemannian "energy integral", a result analogous to the Morse index theorem (see theorem 6.3) holds. More precisely we have:

<u>Lemma 6.10</u> *Let x be a critical point of J. Then, if $\mu(x)$ denotes the geometrical index of x and m(x) is the morse index of x (i.e. the maximal dimension of a subspace on which $J''(x)$ is negative definite), we have $m(x) = \mu(x)$.*

Now we can sketch the proof of theorem 6.6.

Let $z_0 = (x_0, t_0)$ and $z_1 = (x_1, t_1)$ be two nonconjugate points of M. Then, by theorem 6.9, also x_0 and x_1 are nonconjugate (for J). Moreover J satisfies the Palais-Smale compactness condition (see [3]). Hence the Morse relations hold for J, i.e. for every noncritical value $c \in \mathbb{R}$, we have

$$\sum_{x \in Z_J^c} t^{m(x)} = \sum_{k=0}^{+\infty} b_k(c) t^k + (1+t)Q(t),$$

where $J^c = \{x \in \Omega^1 | J(x) \leq c\}$, $Z_J^c = \{x \in J^c | J'(x) = 0\}$, $b_k(c) = \dim H_k(J^c, \mathbb{R})$ and $Q(t)$ is a formal power series with nonnegative integer coefficients. Then, by theorem 6.10, we have

$$\sum_{x \in Z_J^c} t^{\mu(x)} = \sum_{k=0}^{+\infty} b_k(c) t^k + (1+t)Q(t) \qquad (6.4)$$

If in (6.4) we take the limit for $c \longrightarrow +\infty$, we get

$$\sum_{x \in Z_J} t^{\mu(x)} = \sum_{k=0}^{+\infty} b_k(\Omega^1) t^k + (1+t)Q(t), \qquad (6.5)$$

where Z_J is the set of the critical points of J, $b_k(\Omega^1) = \dim H_k(\Omega^1, \mathbb{R})$, and $Q(t)$ is a formal power series with nonnegative integer coefficients (eventually infinite cardinal number).

Now, from Theorem 4.1, Theorem 6.9 (see iii)) and (6.5), we get

$$\sum_{z \in Z_f} t^{\mu(z)} = \sum_{k=0}^{+\infty} b_k(\Omega^1) t^k + (1 + t)Q(t), \qquad (6.6)$$

where Z_f is the set of the critical points of f. Now, since $H^1(t_0, t_1)$ is contractible to a point, Γ^1 and Ω^1 are homotopically equivalent. Hence, for every $k \in \mathbb{N}$, $H_k(\Gamma^1, \mathbb{R}) \cong H_k(\Omega^1, \mathbb{R})$. So, from (6.6), we deduce (6.3).

7. PERIODIC TRAJECTORIES FOR STATIC SPACE-TIMES.

Let us first give the following definition

<u>Definition 7.1</u> Let $M = M_0 \times \mathbb{R}$ be a space-time of split-type and T be a positive real number. Assume that the metric tensor (1.4) is T-periodic in the t variable. A geodesic $z(s) = (x(s), t(s))$ ($s \in [0,1]$) in M is called T-periodic trajectory if it satisfies the following conditions:

$$x(0) = x(1), \quad \dot{x}(0) = \dot{x}(1) \text{ and } t(0) = 0, \quad t(1) = T, \quad \dot{t}(0) = \dot{t}(1)$$

Roughly speaking a T-periodic trajectory is a geodesic whose "spatial" component x is 1-periodic with respect the parameter s and T-periodic with respect the "universal time" t.

We point out that the T-periodic trajectories are not closed geodesics (i.e. geodesics (x(s), t(s)) for which <u>both</u> x(s), t(s) are periodic functions). We refer to [32] for existence results of closed geodesics in stationary Lorentz manifolds.

In this section we shall state a theorem on the existence and multiplicity of the T-periodic trajectories in static space-times.

<u>Theorem 7.2</u> Let $M = M_0 \times \mathbb{R}$ be a static space-time. Assume that M is R-complete (see def.2.1) or a Universe with a black hole (see def.3.1) or a space-time with convex boundary (see def.3.6). Let x_0 be a fixed point in M_0 and d the distance deduced from the Riemannian structure on M_0. Assume that:

i) M_0 is not contractible in itself and $\pi_1(M_0)$ is finite or it has infinitely many coniugacy classes.

ii) The scalar field β (see (1.2)) is bounded, $\beta(x) \longrightarrow \sup \beta$ as $d(x,x_0) \longrightarrow +\infty$ and $\beta(x) < \sup \beta$ for all $x \in M_0$.

iii) There exist a smooth scalar field V on M_0 and positive constants R, ν such that for all $x \in M_0$ with $d(x,x_0) > R$ and for all $\xi \in T_xM_0$ we have

$$H_R^V(x)[\xi,\xi] \geq \nu\langle\xi,\xi\rangle_R.$$

(Here $H_R^V(x)[\xi,\xi]$ denotes the Riemannian hessian of V at x in the direction ξ).

iv) for all $x \in M_0$ with $d(x,x_0) > R$ and for all $\xi \in T_xM_0$ we have:

$$\liminf \langle \nabla_R V(x), \nabla_R \beta(x) \rangle_R = 0 \text{ as } d(x,x_0) \longrightarrow +\infty$$

(∇_R denotes the gradient with respect to the Riemannian metric). Then for any positive integer m there is $T^* > 0$ such that for all $T > T^*$ there are at least m timelike, T-periodic trajectories.

<u>Remark 7.3</u>. If we assume that M_0 is compact then M is R-complete and, of course, we do not need assumptions ii) and iii).

Assumption iii) says that on M_0 a scalar field V, which is convex "at infinity", can be defined. Roughly speaking, this assumption implies that M_0 has "holes" or "handles" only in a bounded region of M_0.

Assumption iv) says that the scalar field β is "sufficiently flat" at infinity.

<u>Remark 7.4.</u> By virtue of the conditions of definition 7.1, it is not difficult to see that the T-periodic trajectories in theorem 7.2 are geometrically distinct (see p. 94,95 in [6]).

<u>Remark 7.5</u> Choosing the function $V = V(r,\vartheta,\phi) = (1/2)r^2$ (see assumption iii)), a straightforward calculation shows that the Schwarzschild space-time satisfies the assumptions of theorem 7.2.

<u>Remark 7.6</u> The results of theorem 7.2 still hold in some cases in which the topology of M_0 is trivial (i.e. assumption i) is not

satisfied). For example existence ([4]) and multiplicity ([6])
results for T-periodic trajectories have been obtained when $M_0 = \mathbb{R}^n$. In this case, in order to get the multiplicity result, we exploit the invariance of the energy functional under the action of the group S^1.

Remark 7.7 Existence results for T-periodic trajectories in "time dependent" space-times of split-type have been obtained in [23,34].

Let us finally observe that the existence of T-periodic trajectories in stationary space-times has been studied in [25,30].

Outline of the proof of theorem 7.2

The proof of theorem 7.2 is completely carried out in [9] only in the case in which M is a universe. It seems clear that the proof is considerably easier when M is R-complete.

In order to avoid some technical difficulties here we shall sketch the proof of theorem 7.2 only in the case in which M is R-complete. This proof can be divided in various steps.

Step 1.

By a slight variant of theorem 4.1, the study of the T-periodic trajectories is reduced to the study of the critical points of the functional

$$J(x) = \int_0^1 \langle \dot{x}(s), \dot{x}(s) \rangle_R \, ds - T^2 \left(\int_0^1 (1/\beta(x)) ds \right)^{-1} \tag{7.1}$$

on the free loop manifold

$\Lambda^1 = \{x : [0,1] \longrightarrow M_0 \text{ with } x \text{ abs. cont.}, \int_0^1 \langle \dot{x}, \dot{x} \rangle_R \, ds < \infty \text{ and } x(0) = x(1)\}$

Since β is bounded the functional J on Λ^1 is bounded from below.

But, since M_0 is not assumed to be compact, J does not satisfy globally the usual Palais-Smale condition. In fact if $\{x_n\}$, $x_n \in M_0$, is a sequence of constant curves with $d(x_0, x_n) \longrightarrow +\infty$, it is easy to see that $J(x_n)$ is bounded and $J'(x_n) \longrightarrow 0$, however $\{x_n\}$ is not precompact. In order to "control" the curves which are "near infinity" we introduce, for $\varepsilon > 0$, the penalized functional

$$J_\varepsilon(x) = \int_0^1 (<\dot{x}(s),\dot{x}(s)>_R \, ds - T^2 \left[\int_0^1 (1/\beta(x))ds\right]^{-1} + \varepsilon \int_0^1 V(x)ds \quad (7.2)$$

Assumption iii) implies that $V(x) \longrightarrow +\infty$ as $d(x_0,x) \longrightarrow +\infty$. This fact easily implies that a sequence $\{x_n\}$ in Λ^1, with $J_\varepsilon(x_n)$ bounded, is bounded in the H^1 norm and then (possibly passing to a subsequence) it weakly converges as $n \longrightarrow +\infty$ to $x_\varepsilon \in \Lambda^1$.

If we assume also that $J'_\varepsilon(x_n) \longrightarrow 0$ as $n \longrightarrow +\infty$, it can be shown, using the tools developed in [5], that $\{x_n\}$ converges (in the H^1 norm) to x_ε. Then we conclude that J_ε satisfies the P.S. condition.

Step 2.

It can be proved that there exists $\delta > 0$ (independent of $\varepsilon > 0$) s.t.

$$\text{cat } (J_\varepsilon^{-T^2+\delta}, \Lambda^1) \leq \dim M_0 \quad \text{for all } \varepsilon > 0. \quad (7.3)$$

where $J_\varepsilon^{-T^2+\delta} = \{x \in \Lambda^1 \mid J_\varepsilon(x) \leq -T^2 + \delta\}$.

In order to prove (7.3), we first recall that

assumption iii) implies that $V(x) \longrightarrow +\infty$ as $d(x_0,x) \longrightarrow +\infty$. Then the set $A = \{x \in \Lambda^1: V(x) \geq V_0\}$, $V_0 > 0$, has compact boundary ∂A. Since β is bounded it is not restrictive to assume $\sup \beta(x) = 1$. Then, by using ii), it can be shown that there exists $\delta = \delta(T) > 0$ (independent on $\varepsilon > 0$) such that

$$J_\varepsilon^{-T^2+\delta} \subseteq \Lambda^1(A) \equiv \{x \in \Lambda^1: x(s) \in A \text{ for all } s \in [0,1]\}. \quad (7.4)$$

By using the gradient flow of the scalar field V we can construct a continuous map

$$\Pi : \Lambda^1(A) \longrightarrow \Lambda^1(\partial A) \text{ s.t. for all } x \in \Lambda^1(A) \quad l(\Pi(x)) \leq kl(x) \quad (7.5)$$

where $\Lambda^1(\partial A) = \{x \in \Lambda^1: x(s) \in \partial A \text{ for all } s \in [0,1]\}$, $l(x)$ denotes the length (with respect the Riemann structure of M_0) of the curve x and $k > 0$ is independent of x.

Now it is easy to see that

for all $x \in J_\varepsilon^{-T^2+\delta}$ $l(x) \leq \delta^{1/2}$ (7.6)

Then, by (7.4) (7.5) and (7.6), we deduce that

$$\Pi(J_\varepsilon^{-T^2+\delta}) \subseteq \Lambda^1(\partial A, \delta) \equiv \{x \in \Lambda^1(\partial A): l(x) \leq k\delta^{1/2}\} \quad (7.7)$$

Since ∂A is compact we can choose δ so small that any curve x in $\Lambda^1(\partial A, \delta)$ has support entirely contained in a geodesically convex neighborhood. Therefore, by a well known procedure (see e.g. [35]), all the curves $x = x(s)$ in $\Lambda^1(\partial A, \delta)$ can be continuously retracted to one of its points (say $x(0)$) along the geodesics joining each point $x(s)$ ($s \in [0,1]$) with $x(0)$.

Therefore by (7.7) and by well known properties of the category (see [39,41]), we get

$$\text{cat}(J_\varepsilon^{-T^2+\delta}, \Lambda^1) \leq \text{cat}(\Lambda^1(\partial A, \delta), \Lambda^1) \leq \text{cat}(\partial A, \Lambda^1) \leq 1 + \dim(\partial A) =$$
$$= \dim M_0.$$

<u>Step 3.</u> By using ii) it can be proved (see [16]) that there exists a sequence $\{K_m\}$ of compact subsets of Λ^1 such that

$$\text{cat}(K_m, \Lambda^1) \longrightarrow +\infty \quad \text{as} \quad m \longrightarrow +\infty \quad (7.8)$$

Then, if m is an integer $> \dim M_0$, we can take $T^* = T^*(m)$ (independent on $0 < \varepsilon \leq 1$) so large that for $T > T^*$

$$J_\varepsilon^{-1} = \{x : J_\varepsilon(x) = \int_0^1 \langle \dot{x}(s), \dot{x}(s) \rangle_R \, ds - T^2 \left(\int_0^1 (1/\beta(x)) ds \right)^{-1} +$$
$$\varepsilon \int_0^1 V(x) ds \leq 1 \} \supseteq K \quad K \text{ compact and with } \text{cat}_{\Lambda^1}(K) \geq m \quad (7.9)$$

By (7.3) and (7.9) there is a "change of topology" between the sublevels J_ε^{-1} and $J_\varepsilon^{-T^2+\delta}$, namely we have

$$\text{cat}(J_\varepsilon^{-1}, \Lambda^1) - \text{cat}(J_\varepsilon^{-T^2+\delta}, \Lambda^1) \geq m - \dim(M_0). \quad (7.10)$$

Since J_ε satisfies the (P.S.) condition (see step 1), by (7.10) and well known arguments of the Lusternik-Schnirelman critical point theory, we deduce that J_ε has at least $m - \dim(M_0)$

critical points x_ε such that

$$-T^2 + \delta \leq J_\varepsilon(x_\varepsilon) \leq -1 \qquad (7.11)$$

The first inequality in (7.11) permits to guarantee that all the critical points x_ε "live" in a bounded set in M_0, more precisely it is possible to prove that there exist $\varepsilon^*, Q > 0$ such that

$$\text{for all } 0 < \varepsilon < \varepsilon^* \text{ and } s \in [0,1] : d(x_0, x_\varepsilon(s)) \leq Q \qquad (7.12)$$

By (7.12) x_ε converges weakly in H^1 as $\varepsilon \longrightarrow 0$ to a curve $x \in \Lambda^1$. Now, passing to the limit $\varepsilon \longrightarrow 0$ in the equation $J'_\varepsilon(x_\varepsilon) = 0$, it can be shown that x solves the equation $J'(x) = 0$.

We observe that in order to preserve the multiplicity of the critical points of J_ε, when we pass to the limit $\varepsilon \longrightarrow 0$, we need to take a different penalization in (7.1).

Let us finally remark that if M is not R-complete (and in this case we assume that M is a universe or that its boundary is convex), we need to add in (7.1) one more penalization term (like in (4.9)) which "prevents" the curves "to touch" the boundary.

8. SOME OTHER PROBLEMS

Here we briefly mention some other existence problems on Lorentzian manifolds which can be studied by suitable variants of the variational methods we have shortly reviewed in this paper.

Harmonic maps.

The paper [26] contains some existence results on the Dirichlet problem for harmonic maps between a Riemannian manifold and a static Lorentzian manifold. We recall that the harmonic map problem reduces to the geodesic problem if the starting manifold is the interval $]0,1[$.

Existence of lightlike geodesics.

A Morse theory for lightlike geodesics joining a point with a timelike curve has been developed in [42] (see also sec. 9.3 of [2]).) for globally hyperbolic space-times. Existence and multiplicity results were obtained in [14, 17] for static and stationary Lorentzian manifolds with boundary. In [17] a Morse theory for lightlike geodesics in stationary space-times with boundary is developed.

Uniqueness of geodesics.

The existence of a unique geodesic joining two given events in a static space-time has been studied in [18].

REFERENCES

[1] Avez A. *Essais de geometrie Riemannienne hyperbolique globale. Application a la Relativite' Generale*, Ann. Inst. Fourier.**132**,(1963), 105-190.

[2] Beem J.K., Herlich E.: *Global Lorentzian Geometry.* New York-Basel, Marcel Dekker Inc. 1981.

[3] Benci V.,Fortunato D. *Existence of geodesics for the Lorentz metric of a stationary gravitational field.* Ann. Inst. H. Poincare'. Analyse nonlineaire.**7**,(1990),27-35.

[4] Benci V.,Fortunato D. *Periodic trajectories for the Lorentz metric of a static gravitational field.* Proc. ''Variational Problems'', Paris, June 1988, H. Berestycki- I. Ekeland- M. Coron editors. Birkhauser.

[5] Benci V.,Fortunato D. *On the existence of infinitely many geodesics on space-time manifolds.* Adv. Math. to appear.

[6] Benci V.,Fortunato D.,Giannoni F.*On the existence of multiple geodesics in static space-time.* Ann. Inst. H.Poincare'. Analyse nonlineaire. **8** ,(1991),79-102.

[7] Benci V.,Fortunato D.,Giannoni F. *On the existence of geodesics in static Lorentz manifolds with singular boundary.* Ann. Scuola Norm. Sup. Pisa. **19** , (1992).

[8] Benci V.,Fortunato D.,Giannoni F.*Geodesics on static Lorentz manifolds with convex boundary.* Proc. ''Variational methods in

Hamiltonian systems and elliptic equations''. M. Girardi, M. Matzeu, F. Pacella editors. L'Aquila, January 1990. Pitman.

[9] Benci V.,Fortunato D.,Giannoni F. *On the existence of periodic trajectories in static Lorentz manifolds with non smooth boundary.* Nonlinear Analysis, a tribute in honor of G. Prodi- Quaderni Scuola. Norm. Sup. Pisa, Ambrosetti A. and Marino A. editors (1991),109-133.

[10] Benci V., Fortunato D., Masiello A., in preparation.

[11] Benci V., Masiello A., *A Morse index for geodesics in static Lorentz manifolds,* Math. Annalen, **293**, (1992), 433-442.

[12] Benci V., Fortunato D., Masiello A., *On the geodesic connectedeness of Lorentzian manifolds,* preprint Univ. Bari (1992).

[13] Bott R., *Lectures on Morse theory old and new.* Bull. Am. Math. Soc.,**6**,(1982),331-358.

[14] Capozzi A., Fortunato D., Greco C., *Null geodesics on Lorentz manifolds,* Preprint, Univ. Bari,(1992).

[15] Faddell E., Husseini A., *Category of loop spaces of open subsets in euclidean space,* Nonlinear Anal. Theory,Meth.,Appl.,**17**, 1153-1161, (1991).

[16] Faddell E., Husseini, *Infinite cup length in free loop spaces, with an application to a problem on the n-body type,* Ann. Inst. H. Poincare'. Analyse nonlineaire,**9**,(1992),305-319.

[17] Fortunato D., Giannoni F., Masiello A., *Lightlike geodesics in stationary lorentzian manifolds,* in preparation.

[18] Frattarolo W.,*A Hadamard-type theorem for the static space-time.* Quaderno n.4 Ist. Mat.Appl. Pisa (1991).

[19] Geroch R.,*Domains of dependence.* J.Math.Phys.**11**,(970),437-449

[20] Giannoni F., *Geodesics on non static Lorentz manifolds of Reissner-Nordstrom type,* Math. Ann. **291**, 383-401 (1991).

[21] Giannoni F.,Masiello A., *On the existence of geodesics in stationary Lorentz manifolds with convex boundary,* J. of Funct.

Anal. **101**, 340-369, (1991).

[22] Giannoni F., Masiello A., *Geodesics on Lorentzian manifolds with quasi-convex boundary,* preprint.

[23] Greco C., *Periodic trajectories for a class of Lorentz metrics of a time-dependent gravitational field,* Math. Ann. **287**, 515-521 (1990).

[24] Greco C., *Periodic trajectories in static space-times,* Proc. of Royal Soc. of Edin. **113A**, 99-103 (1989)

[25] Greco C., *Multiple periodic trajectories on stationary space-times,* Ann. Mat. Pura e Appl., to appear.

[26] Greco C., *The Dirichlet problem for harmonic maps from the disk into a Lorentzian warped product,* Ann. Inst. H. Poincare'. Analyse nonlineaire, to appear.

[27] Greco C., *Infinitely many periodic trajectories on a class of static Lorentzian manifold.* Preprint.

[28] Hawking S. W., Ellis G. F.R., *The large scale structure of space-time.* Cambridge University Press (1973)

[29] Klingenberg W., *Riemannian Geometry.* Berlin-New York Walter de Gruiter (1982).

[30] Masiello A., *Timelike periodic trajectories on stationary Lorentz manifolds,* Nonlinear Analysis T.M.A., to appear.

[31] Masiello A., *Some results on the geodesic connectedness of lorentzian manifolds.* This volume.

[32] Masiello A., *On the existence of a closed geodesic in stationary Lorentz manifolds,* J. Diff. Eq., to appear.

[33] Masiello A., *Metodi variazionali in geometria lorenziana.* Tesi Dott. Ricerca. Univ. Pisa.

[34] Masiello A., Pisani L., *Existence of a timelike periodic trajectory for a time dependent Lorentz metric,* Ann. Univ. Ferrara, to appear.

[35] Milnor J.,*Morse Theory.* Ann. Math. Studies **51**,(1963),Princeton University Press.

[36] O' Neill, *Semiriemannian Geometry with application to Relativity,* Academic Press Inc., New York-London, 1983.

[37] Penrose R., *Techniques of differential topology in Relativity,* Conf. Board Math. Sci. **7**,(1972),S.I.A.M.

[38] Pisani L., *Existence of geodesics for stationary Lorentz manifolds,* Boll. U.M.I. (7) **5-B**, 507-520 (1991).

[39] Schwartz J.,*Nonlinear functional analysis,* Gordon and Breach, New York, 1969.

[40] H.J.Seifert, *Global connectivity by time-like geodesics,* Z. Naturefor. 22a, (1970), 1356.

[41] Struwe M., *Variational Methods: applications to nonlinear partial differential equations and Hamiltonian systems,* Springer-Verlag, Berlin, Heidelberg, New-York, 1990.

[42] Ulenbeck K., *A Morse theory for geodesics on a Lorentz manifold,* Topology **14**, 69-90, 1975

THE ROLE OF THE MEAN CURVATURE IN SOME NONLINEAR PROBLEM WITH CRITICAL EXPONENT

ADIMURTHI[*], FILOMENA PACELLA[**] and S.L. YADAVA[*]

(*) T.I.F.R. Centre, P.B. 1234, Bangalore 560012, India
(**) Dipartimento di Matematica – Università di Roma
"La Sapienza" – P.le A. Moro 2 – 00185 Roma, Italia

1. INTRODUCTION

In this paper we consider the equation

$$-\Delta u + \lambda u = |u|^{p-1} u \quad \text{in } \Omega \tag{1.1}$$

with the boundary condition

$$\frac{\partial u}{\partial \nu} = 0 \quad \text{on } \partial\Omega \tag{1.2}$$

or

$$u = 0 \quad \text{on } \Gamma_0 \quad \text{and} \quad \frac{\partial u}{\partial \nu} = 0 \quad \text{on } \Gamma_1. \tag{1.3}$$

In (1.1)–(1.3) Ω is a smooth bounded domain in $\mathbb{R}^N$, $N \geq 3$, Γ_0 and Γ_1 are two disjoint components of $\partial\Omega$ such that $\partial\Omega = \overline{\Gamma}_0 \cup \overline{\Gamma}_1$, ν denotes the outward normal to $\partial\Omega$ and p and λ are real numbers, $1 < p$.

The equation (1.1) with the Neumann boundary condition (1.2) arises naturally in the study of mathematical models in biological pattern formation theory. One of these models is the chemotactic aggregation model, due to Keller and Segal[13,15], of cellular slime molds which release a certain chemical and move toward places of its higher concentration, forming aggregates. The study of (1.1)–(1.2) is also relevant for an activator–inhibitor type system due to Gierer–Meinhardt (see [15]). Various numerical studies have been done and one of the most interesting features expected from numerical simulations is that the solutions of (1.1)–(1.2) seem to exhibit "point–condensation" phenomena, i.e. they tend to zero, as $\lambda \to \infty$, except at a finite number of points. Therefore it is important to study not only the existence or multiplicity of solutions of (1.1), but also the shape of the solutions. In this paper we review some recent results which emphasize the role of the mean curvature of $\partial \Omega$ with respect to these kind of questions. We will focus our exposition on the critical case $p = \frac{N+2}{N-2}$ and will give some information about the subcritical $\left(p < \frac{N+2}{N-2}\right)$ or supercritical $\left(p > \frac{N+2}{N-2}\right)$ case at the end.

The paper is organized as follows. In Section 2 we describe some existence and multiplicity results while Section 3 is dedicated to some geometrical properties of solutions with low energy. In the final part of Section 3 we will mention a few results about the subcritical and supercritical case.

2. EXISTENCE AND MULTIPLICITY OF POSITIVE SOLUTIONS

Here we are concerned with the existence and multiplicity of positive solutions of (1.1) in the space $H^1(\Omega)$, in the case of the Neumann condition (1.2) or in the space $H^1(\Gamma_0) = \{u \in H^1(\Omega) \text{ such that } u = 0 \text{ on } \Gamma_0\}$ in the case of the mixed boundary condition (1.3). Let us remark immediately that a necessary condition for the existence of positive solutions is $\lambda > -\lambda_1(\Omega)$ where $\lambda_1(\Omega)$ is the first eigenvalue of $-\Delta$ in the space $H^1(\Gamma_0)$; hence λ must be positive if we are considering the Neumann problem. In this last case it is easy to see that the function $w_\lambda(x) = \lambda^{\frac{1}{p-1}}$ is a positive solution of (1.1)–(1.2) and is the only constant solution.

Since the mixed boundary problem is more general than the Neumann problem, in the sense that (1.3) reduces to (1.2) when $\Gamma_0 = \emptyset$, we will mainly consider this case.

One way of finding a positive solution of (1.1), (1.3) is to minimize the functional

$$Q_\lambda(u) = \frac{\int_\Omega |\nabla u|^2 + \lambda \int_\Omega |u|^2}{\left(\int_\Omega |u|^{2*}\right)^{2/2*}}, \quad 2^* = p + 1 = \frac{2N}{N-2} \qquad (2.1)$$

in the space $H^1(\Gamma_0)\setminus\{0\}$, i.e. to prove that the infimum

$$S_\lambda = \inf_{H^1(\Gamma_0)\setminus\{0\}} Q_\lambda(u) \qquad (2.2)$$

is achieved.

A first result in this direction was obtained by Lions, Pacella and Tricarico who gave some examples of domains for which S_λ is achieved, in the case $\lambda = 0$ (see [17], Remark 2.1).

Subsequently a much more complete result was obtained, at the same time, by Adimurthi and Mancini[1] and Wang[21]. Their result is the following.

THEOREM 2.1. *If there exists a point $x_0 \in \Gamma_1$ such that the mean curvature $H(x_0)$, at x_0, is positive, then S_λ is achieved.*

Since Ω is bounded from this theorem we immediately deduce

COROLLARY 2.1. *The Neumann problem (1.1)–(1.2) has always a solution u_λ which minimizes Q_λ.*

Remark 2.1. Note that since the constant solution $w_\lambda(x) = \lambda^{\frac{1}{p-1}}$ always exists it is important to find out whether $u_\lambda = w_\lambda$ or not. By means of an estimate of $Q_\lambda(w_\lambda)$ it is possible to prove that $u_\lambda \neq w_\lambda$ for λ sufficiently large (see [1,21]).

Let us sketch the proof of Theorem 2.1. By the concentration–compactness principle of Lions[16] (see also [17]) it is possible to prove that a minimizing sequence u_n either is relatively compact or concentrates in a point of Γ_1 (i.e. $|u_n|^{2*} \rightharpoonup k\delta_z$ for some $z \in \Gamma_1$, $k > 0$). In the last case $Q_\lambda(u_n) \to \Sigma$ where Σ is the best Sobolev constant for the embedding $D^1(\mathbf{R}_+^N) \hookrightarrow L^{2*}(\mathbf{R}_+^N)$, with $D^1(\mathbf{R}_+^N) = \{u \in L^{2*}(\mathbf{R}^N) : |\nabla u| \in L^2(\mathbf{R}^N)\}$. Therefore to avoid the concentration phenomenum it is sufficient to prove that $S_\lambda < \Sigma$. To do that let us introduce the functions

$$U_{\varepsilon,y}(x) = \frac{1}{\varepsilon^{\frac{N-2}{2}}} U\left(\frac{x-y}{\varepsilon}\right) \tag{2.3}$$

$\varepsilon > 0$, $y \in \mathbf{R}^N$ and $U(x) = \left[\frac{N(N-2)}{N(N-2)+|x|^2}\right]^{\frac{N-2}{2}}$.

The following lemma holds

LEMMA 2.1. *There exist positive constants A_N and a_N such that for all $\varepsilon > 0$, $y \in \partial\Omega$ and $\lambda \geq 1$*

$$Q_\lambda(U_{\varepsilon,y}) = \Sigma - A_N H(y)\beta_1(\varepsilon) + a_N \lambda \beta_2(\varepsilon) + O(\beta_2(\varepsilon)) + o(\lambda\beta_2(\varepsilon)) \tag{2.4}$$

where

$$\beta_1(t) = \begin{cases} t\log\frac{1}{t} & \text{if } N = 3 \\ t & \text{if } N \geq 4 \end{cases}$$

$$\beta_2(t) = \begin{cases} t & \text{if } N = 3 \\ t^2 \log\frac{1}{t} & \text{if } N = 4 \\ t^2 & \text{if } N \geq 5 \end{cases} \qquad \beta_3(t) = \begin{cases} t^{1/2} & \text{if } N = 3 \\ t\left(\log\frac{1}{t}\right)^{2/3} & \text{if } N = 4 \\ t & \text{if } N \geq 5 \end{cases}$$

Proof. See the proof of (2.10) of Lemma 2.2 in [1]. Thus, considering the function $V_{\varepsilon,x_0}(x) = \varphi(x)U_{\varepsilon,x_0}(x)$, where $\varphi(x)$ is a suitable cut function chosen to satisfy the boundary condition on Γ_0, and using (2.4) we get that $S_\lambda < \Sigma$, since $H(x_0) > 0$.

So far we have shown how the positiveness of the mean curvature in a point of Γ_1 produces a solution which is actually a function which minimizes Q_λ in $H^1(\Gamma_0)$; from now on such a solution will be called a least-energy solution. Therefore it is natural to ask whether the assumption on the mean curvature is necessary or not to achieve S_λ. In other word the question is: suppose that $H(y) \leq 0$, for any $y \in \Gamma_1$, is then the infimum achieved?

In a paper by Egnell, Tricarico and Pacella ([10,20]), as a consequence of certain Sobolev inequalities with lower order terms in the space $H^1(\Gamma_0)$ it is proved that if Ω is a bounded domain with a spherical hole and if Γ_1 is the boundary of this hole then the infimum is Σ and it is not achieved for any $\lambda \geq 0$. This seems to lead to the conjecture that S_λ is not achieved when the mean curvature is negative (or zero) at any point of Γ_1 and Γ_1 is made only by one component.

Of course when the infimum is achieved one could investigate the existence of positive solutions of (1.1), (1.3) by other methods. This has been done in[11] for some domains with two holes and then generalized in[2] and again in[8].

Now we would like to state a multiplicity result obtained in[3,4] which also emphasizes the role of the mean curvature in these kind of problems.

For simplicity we will consider only the case of the Neumann condition (1.2), referring the reader to[4] for the mixed boundary problem.

THEOREM 2.2. *Let $N \geq 7$ and $P_0 \in \partial\Omega$ be a strict local maximum point of $H(x)$, such that $H(P_0) > 0$. Then there exists a $\lambda_0 > 0$ such that for all $\lambda > \lambda_0$, problem (1.1)–(1.2) has a solution u_λ with $Q_\lambda(u_\lambda) < \Sigma$. Moreover the functions u_λ concentrate at P_0, as $\lambda \to \infty$.*

Of course from this theorem we deduce the existence of k positive solutions of (1.7) in any domain Ω with k "peaks", i.e. with k points on the boundary of strict local maximum for the mean curvature $H(x)$.

The crucial point in the proof of Theorem 2.2 is the following lemma

LEMMA 2.2. *Let $\delta_k > 0$, $\lambda_k > 0$, $x_k \in \partial\Omega$ and $0 \leq u_k \in H^1(\Omega)$ be such that as $k \to \infty$, $\lambda_k \to \infty$, $x_k \to x_1$, $\delta_k \to 0$, $u_k \to 0$ weakly in $H^1(\Omega)$ and*

$$\lim_{k \to \infty} \int_\Omega |\nabla(u_k - U_{\delta_k,x_k})|^2 = 0. \qquad (2.5)$$

Then there exist $\varepsilon_k > 0$, $C_k \in \mathbf{R}$, $y_k \in \partial\Omega$, $\omega_k \in H^1(\Omega)$ such that for a subsequence, as $k \to \infty$, $\varepsilon_k/\delta_k \to 1$, $C_k \to 1$, $y_k \to P_0$ and

$$u_k = C_k U_{\varepsilon_k, y_k} + \omega_k .$$

Moreover, if $N \geq 7$ and $a_{\lambda_k}(u_k) < \Sigma$, there exists $r > 2$ such that

$$Q_{\lambda_k}(u_k) \geq \Sigma - A_N H(y_k)\beta_1(\varepsilon_k) + a_N \lambda_k \beta_2(\varepsilon_k) + o(\lambda_k \beta_2(\varepsilon_k)) + \\ + O(\beta_2(\varepsilon_k) + \beta_3(\varepsilon_k)\beta_1(\varepsilon_k)^{1/2} + \beta_1(\varepsilon_k)^{r/2}) \qquad (2.6)$$

$$\lambda_k \beta_2(\varepsilon_k) = O(\beta_1(\varepsilon_k)) \qquad (2.7)$$

where the functions β_i are as defined in Lemma 2.1. Moreover if u_k solves (1.1)-(1.2) then

$$Q_{\lambda_k}(u_k) = \Sigma - A_N H(y_k)\beta_1(\varepsilon_k) + a_N \lambda_k \beta_2(\varepsilon_k) + o(\lambda_k \beta_2(\varepsilon_k)) \\ + O(\beta_2(\varepsilon_k) + \beta_3(\varepsilon_k)^2) \qquad (2.8)$$

For the proof of this lemma we refer to[4]. We now sketch the proof of Theorem 2.2. By the hypothesis there exists $R_0 > 0$ such that for all $x \in \partial\Omega \cap \overline{B(P_0, R_0)}\setminus\{P_0\}$

$$0 \leq H(x) < H(P_0) \qquad (2.9)$$

where $B(P_0, R_0)$ is the ball of center P_0 and radius $R_0 > 0$. Let us define

$$A = \left\{u \in H^1(\Omega) : \rho(u) \in \overline{B(P_0, R_0)}\right\}, \quad \rho(u) = \frac{\int_\Omega x|\nabla u|^2}{\int_\Omega |\nabla u|^2} \qquad (2.10)$$

and

$$S_\lambda(P_0) = \inf\{Q_\lambda(u) : u \in A\setminus\{0\}\} . \qquad (2.11)$$

It is easy to see that A is not empty since the functions U_{ε, P_0} of (2.3) belong to A for ε sufficiently small. Moreover, by (2.4) $S_\lambda(P_0) < \Sigma$ and therefore $S_\lambda(P_0)$ is achieved by a function u_λ that we can assume greater or equal to zero (otherwise it can be replaced by $|u_\lambda|$).

In order to show that u_λ, for large λ, are required solutions, it is sufficient to prove that there exists $\lambda_0 > 0$ such that for all $\lambda > \lambda_0$, $\rho(u_\lambda) \in B(P_0, R_0)$.

Suppose that this is not true. Then there exist sequences $\lambda_k \to \infty$, $u_k = u_{\lambda_k}$ such that $\rho(u_k) \in \partial B(P_0, R_0)$. Furthermore it is possible to prove (see [4]) that u_k satisfies the hypotheses of Lemma 2.2 with $x_k \to x_1 \in \partial\Omega \cap \partial B(P_0, R_0)$. Thus there exist sequences $\varepsilon_k \to 0$, $y_k \in \partial\Omega$ with

$y_k \to x_1$ such that $Q_{\lambda_k}(u_k)$ satisfies (2.6). From (2.4), (2.6), (2.7) and $Q_{\lambda_k}(u_k) \leq Q_{\lambda_k}(U_{\varepsilon_k,P_0})$ we get

$$A_N(H(P_0) - H(y_k))\beta_1(\varepsilon_k) + o(\beta_1(\varepsilon_k)) \leq 0. \qquad (2.12)$$

On the other side, from (2.9) we have $\lim_{k\to\infty}(H(P_0) - H(y_k)) = H(P_0) - H(x_1) > 0$ which contradicts (2.12) and hence proves that $\rho(u_\lambda) \in B(P_0, R_0)$ for λ sufficiently large.

We conclude this section remarking that another multiplicity result for (1.1)–(1.2) which also exploits the mean curvature of $\partial\Omega$ has recently been obtained by Adimurthi and Mancini[2] and Wang[22] at the same time.

3. SHAPE OF SOLUTIONS WITH LOW ENERGY

Once we have found some positive solutions of (1.1)–(1.2), for example with the methods described in the previous section, it is natural to ask questions about the qualitative behaviour of the solutions or about their geometrical properties. Where in $\overline{\Omega}$ does a solution achieve its maximum? At how many points? If Ω is a symmetric domain does a positive solution share the same symmetry?

In this section we would like to emphasize the role of the mean curvature also in answering this kind of questions. A first result obtained in[4] is the following

THEOREM 3.1. *Let $u_\lambda \in H^1(\Omega)$ be a solution of (1.1)–(1.2) such that $Q_\lambda(u_\lambda) < \Sigma$. Then there exists $\lambda_0 > 0$ such that, for all $\lambda > \lambda_0$*
  a) *u_λ attains its maximum at only one point $P_\lambda \in \partial\Omega$*
  b) *further if $N \geq 7$ and u_λ is a least–energy solution (i.e. $Q_\lambda(u_\lambda) = S_\lambda$) then the limit points of $\{P_\lambda\}$, as $\lambda \to \infty$, are contained in the set of the points of maximum mean curvature.*

Proof. The method of the proof of a) is similar to the one used by Ni and Takagi[19] for the subcritical case and relies on a blow up technique used in Gidas and Spruck[11]. We refer the reader to[4] for the details. To prove b) let u_λ be a least energy solution of (1.1)–(1.2) and suppose that for a sequence $\lambda_k \to \infty$, $P_{\lambda_k} \to P_0$. Then it is possible to prove (see Lemma 2.2 of [4]) that $u_k = u_{\lambda_k}$, $x_k = P_{\lambda_k}$ and $\delta_k = (u_k(P_k))^{-2/N-2}$ satisfy the hypotheses of Lemma 2.2 and hence there exist ε_k, y_k satisfying (2.7), (2.8). Suppose that P_0 is not a point of maximum mean curvature and let y_0 be a point of maximum mean curvature. Then, since $Q_{\lambda_k}(u_k) \leq Q_{\lambda_k}(U_{\varepsilon_k,y_0})$ it follows from (2.7), (2.8) and (2.4) that

$$A_N(H(y_0) - H(y_k))\beta_1(\varepsilon_k) + o(\beta_1(\varepsilon_k)) \leq 0$$

which is a contradiction. This proves the theorem.

We would like to point out that a result similar to a) of Theorem 3.1 has been obtained in[18].

So far we have studied the behaviour, as $\lambda \to \infty$, of positive solutions which blow up at points on $\partial\Omega$ which are either maximum points or local maximum points for the mean curvature $H(x)$, (mainly part b) of Theorem 3.1 and Theorem 2.2). On the other side statement a) of Theorem 3.1 holds for any positive solution u_λ with energy smaller than Σ. Then it is natural to ask whether one can characterize completely the concentration points of any positive solution u_λ with $Q_\lambda(u_x) < \Sigma$.

Recently we have answered this question with the following

THEOREM 3.2. *Let $N \geq 7$, u_λ be a positive solution of (1.1)–(1.2) with $Q_\lambda(u_\lambda) < \Sigma$ and $P_\lambda \in \partial\Omega$ its unique maximum point. Then there exists a $\lambda_0 > 0$ such that for $\lambda > \lambda_0$, the limit points of P_λ, as $\lambda \to \infty$, are contained in the set of critical points of H.*

The proof of this theorem relies on Lemma 2.2 and other, rather long, estimates for which we refer the reader to[5]. From similar estimates it is also possible to deduce some information on the rate of blow-up of the solutions. More precisely we have

THEOREM 3.3. *let $N \geq 7$ and u_λ, P_λ be as in the previous theorem. If $\lambda_k \to \infty$, $u_k = u_{\lambda_k}$ and $P_{\lambda_k} \to P_0$, then*

$$\lim_{k \to \infty} \frac{\|u_k\|_\infty^{2/N-2}}{\lambda_k} = \left[\frac{\omega_N N^{1/2}(N-1)\Gamma(N/2)\Gamma\left(\frac{N-4}{2}\right)}{\omega_{N-1}(N-2)^{1/2}\Gamma\left(\frac{N+1}{2}\right)\Gamma\left(\frac{N-3}{2}\right)}\right]\frac{1}{H(P_0)}$$

$$\lim_{k \to \infty} \lambda_k(Q_{\lambda_k}(u_k) - \Sigma) = \frac{(N-1)\Sigma\Gamma\left(\frac{N-3}{2}\right)^2}{\pi N(N-4)\Gamma\left(\frac{N-4}{2}\right)^2}H(P_0)^2$$

where $\|u_k\|_\infty$ denotes the L^∞-norm of u_k and ω_N is the volume of the unit sphere in $\mathbb{R}^N$.

Proof. See [5].

Remark 3.1. All positive solutions u_λ of (1.1)–(1.2) considered throughout this paper are solutions with low energy (i.e. $Q_\lambda(u_\lambda) < \Sigma$) and we have shown that they exhibit only one peak, as $\lambda \to \infty$. It would be interesting to find solutions with high energy, especially multi-peak solutions.

We close this section mentioning some results for problem (1.1)–(1.2) with $p \neq \frac{N+2}{N-2}$. One of the first results for the subcritical case $\left(p < \frac{N+2}{N-2}\right)$ was obtained by Lin, Ni and Takagi[15] and can be summarized in the following

THEOREM 3.4. *Let $p < \frac{N+2}{N-2}$. Then any positive solution of (1.1)–(1.2) is constant if λ is sufficiently small. Moreover there exists a positive number $\lambda_0 > 0$ such that, for any $\lambda > \lambda_0$, (1.1)–(1.2) admits a nonconstant positive solution.*

An estimate of the number λ_0 is given in the paper[14]. In the same article a partial result for the supercritical case is shown.

THEOREM 3.5. *Let $p > \frac{N+2}{N-2}$ and Ω be a ball. Then there exists $\tilde{\lambda} > 0$ such that (1.1)–(1.2) admits a positive radial non constant solution for any $\lambda > \tilde{\lambda}$ while does not have any positive radial nonconstant solution for $\lambda < \tilde{\lambda}$.*

The last two theorems seem to lead to the conjecture that for any $p > 1$ and λ small do not exist positive nonconstant solutions to (1.1)–(1.2). It is very interesting to note that this is not the case for $p = \frac{N+2}{N-2}$ in virtue of the following result due to Adimurthi and Yadava[6] and Budd, Knaap and Peletier[9].

THEOREM 3.6. *Let Ω be a ball in $\mathbf{R}^N$, $N = 4, 5, 6$. Then there exists $\tilde{\lambda} > 0$ such that for $\lambda < \tilde{\lambda}$ (1.1)–(1.2) has nonconstant positive radial solutions.*

The previous three theorems emphasize the structural differences between the critical case and the other cases. Moreover in Theorem 3.6, as well as in Theorem 2.2, 3.1, 3.2. there are hypotheses on the dimension N which are, at a first glance, rather strange. People who are familiar with nonlinear equations with "critical exponents" know that these kind of hypotheses often appear, also in Dirichlet–type problems. It is an interesting open question to understand the role played by the dimension in these type of nonlinear problems.

REFERENCES

1. Adimurthi and G. Mancini, Scuola Norm. Sup. Pisa, volume in honour of G. Prodi (1991).
2. Adimurthi and G. Mancini, (to appear).
3. Adimurthi, F. Pacella and S.L. Yadava, C.R. Acad. Sc. Paris (1992).
4. Adimurthi, F. Pacella and S.L. Yadava, Journ. Funct. Anal. (to appear).
5. Adimurthi, F. Pacella and S.L. Yadava, (to appear).
6. Adimurthi and S.L. Yadava, Arch. Rat. Mech. Anal. (to appear).
7. Adimurthi and S.L. Yadava, Proc. Ind. Acad. Sc., 100 (1990).
8. H. Berestycki, M. Grossi and F. Pacella, (to appear).
9. G. Budd, M.C. Knaap and L.A. Peletier, Proc. Royal Soc. Edinb. (to appear).

10. H. Egnell, F. Pacella and M. Tricarico, Nonlin. Anal. T.M.A. (1989).
11. B. Gidas and J. Spruck, Comm. Part. Diff. Eq. 6 (1981).
12. M. Grossi and F. Pacella, Proceed. Royal Soc. Edinburgh 116A (1990).
13. E. Keller and L. Segal, J. Theor. Biol. 26 (1970).
14. C.S. Lin and W.M. Ni, Calc. of Variations and Part. Diff. Eq., Lecture Notes in Math. 1340, Springer–Verlag (1988).
15. C.S. Lin, W.M. Ni and I. Takagi, J. Diff. Eq., 72 (1988).
16. P.L. Lions, Riv. Mat. Ibeoramericana 1 (1985).
17. P.L. Lions, F. Pacella and M. Tricarico, Indiana Univ. Math. Journ. 37 (1988).
18. W.M. Ni, X.B. Pan and I. Takagi, (to appear).
19. W.M. Ni and I. Takagi, Comm. Pure Appl. Math., XLIV (1991).
20. F. Pacella, Progress Nonlin. Diff. Eq., vol. 4, Birkhäuser (1990).
21. X.J. Wang, J. Diff. Eq. 93 (1991).
22. Z.Q. Wang, (to appear).

NONTRIVIAL SOLUTIONS TO SOME NONLINEAR EQUATIONS VIA MINIMIZATION

D. ARCOYA[*] and L.BOCCARDO[**]

(*) Universidad de Granada
(**) Universitá di Roma 1

1. INTRODUCTION

We shall be concerned with a classical minimization problem of the calculus of variations which is to find the functions which minimize a given functional. Many papers deal with functionals of the type

$$\tfrac{1}{2}\int_\Omega |Dv|^2 - \tfrac{1}{q}\int |v|^q \;,\quad q\in(1,2).$$

A small change in the principal part can give rise an important change in the Euler equation. We shall study functionals whose simplest example is

$$\tfrac{1}{2}\int_\Omega a(x,v)|Dv|^2 - \tfrac{1}{q}\int_\Omega |v|^q \;,\quad q\in(1,2).$$

2. MINIMIZATION PROBLEM

Consider the functional

(2.1) $$I(v) = \int_\Omega f(x,v,Dv) \;,\quad v\in W_0^{1,p}(\Omega)$$

where Ω is a bounded, open subset of $\mathbb{R}^N$, $p>1$ and $f(x,s,\xi)$ a Caratheodory function such that

(2.2) $\quad\quad\quad\quad\quad\quad \xi \to f(x,s,\xi) \quad$ is convex

(2.3) $\quad\quad\quad\quad\quad \beta|\xi|^p \geq f(x,s,\xi) \geq \alpha|\xi|^p \quad,\quad$ for some $\alpha>0$, $\beta>0$.

We are interested in the nontrivial solutions of the minimization problem

(2.4) $\quad\quad\quad \min\left\{J(v) = I(v) - \frac{1}{q}\int_\Omega (v^+)^q : v \in W_0^{1,p}(\Omega), 1<q<p\right\}$

<u>Theorem 1</u> - Under the assumptions (2.2), (2.3) there exists $u \geq 0$, $u \neq 0$, solution of the minimization problem (2.4). Moreover $u \in L^\infty(\Omega)$.

Proof. - The assumptions (2.2), (2.3) and $q<p$ guarantee the existence of u. Let φ the first eigenfunction of the p-laplacian. Then

$$I(\epsilon\varphi) - \frac{1}{q}\int_\Omega |\epsilon\varphi|^q \leq \epsilon^p\beta\int |D\varphi|^p - \frac{1}{q}\epsilon^q\int\varphi^q = \epsilon^p\beta\lambda_1\int\varphi^p - \frac{1}{q}\epsilon^q\int\varphi^q < 0 = I(0) - \frac{1}{q}\int 0^q$$

for ϵ small enough.

Moreover $u \geq 0$ is a consequence of the definition of minimum. Indeed $J(u) \leq J(u^+)$ implies

$$\alpha \int_{\{u<0\}} |Du|^p < 0.$$

Finally we shall prove that u belongs to $L^\infty(\Omega)$. We shall follow the Stampacchia method. Fix $k>0$. For any $s \in \mathbb{R}$, define

$$T_k(s) = \begin{cases} s & \text{if } |s| \leq k \\ \frac{ks}{|s|} & \text{if } |s| > k. \end{cases}$$

and

$$G_k(s) = s - T_k(s)$$

According to the definition of minimum

$$J(u) \leq J(T_k(u)).$$

Recall that $DG_k(u) = Du$ on the set

$$A(k) = \{x \in \Omega : |u(x)| > k\}$$

whereas $Du = 0$ elsewhere ([S]). Thus

$$\int_{A(k)} f(x,u,Du) - \frac{1}{q}\int_\Omega u^q \leq -\frac{1}{q}\int_\Omega T_k(u)^q$$

and

$$\int_\Omega |DG_k(u)|^p \leq c_1 \int_{A(k)} (u^q - k^q).$$

Sobolev's inequality and Lipschitz continuity of the real function t^q imply

$$\left[\int_\Omega |G_k(u)|^{p^*}\right]^{p/p^*} \leq c_2 \int_{A(k)} (u^q - k^q) \leq$$

$$= c_2 \int_{A(k)} G_k(u) \frac{u^q - k^q}{u-k} \leq c_3 \int_{A(k)} G_k(u) u^{q-1}$$

If $h > k > 0$, $A(h) \subset A(k)$ and we have that

$$(h-k)^{p-1} |A(h)|^{\frac{p-1}{p^*}} \leq c_4 \left[\int_{A(k)} u^{(q-1)p^*1}\right]^{1/p^*1}$$

Now we use the results of [S] to the derive that u belongs to $L^m(\Omega)$, $m > p^*$ then using again [S] we shall prove that u is bounded.

3. EULER EQUATION

Roughly speaking it is well known (see [LU] e.g.) that $< J'(u), v >$ meakes sense for $v \in W_0^{1,p}(\Omega) \cap L^\infty$, that is: $u \in W_0^{1,p}(\Omega)$ satisfies the weak form of the Euler equation

(3.1) $$\int_\Omega a(x,u,Du)Dv + \int_\Omega b(x,u,Du)v = \int_\Omega |u|^{q-2} u\, v$$

for any $v \in W_0^{1,p}(\Omega) \cap L^\infty(\Omega)$, where

$$a(x,s,\xi) = \frac{\partial f}{\partial \xi}$$

$$b(x,s,\xi) = \frac{\partial f}{\partial s}.$$

Assume that a and b are Caratheodory function such that

(3.2) $$a(x,s,\xi)\xi \geq \alpha|\xi|^p, \text{ for some } \alpha > 0$$

(3.3) $|a(x,s,\xi)| \leq \beta\left[k)x\right) + |s|^{p-1} + |\xi|^{p-1}\right]$, for some $\beta > 0$ and $k(x) \in L^{p'}(\Omega)$

(3.4) $$\left[a(x,s,\xi) - a(x,s,\xi^*)\right]\left[\xi - \xi^*\right] > 0, \quad \xi \neq \xi^*$$

(3.5) $$|b(x,s,\xi)| \leq h(x) + \gamma |\xi|^p, \quad h \in L^1(\Omega),\, \gamma > 0$$

(3.6) $$sb(x,s,\xi) \geq 0.$$

Then the Euler equations makes sense.

We point out we are not able to discuss the existence of other critical points: we cannot use the Mountain-pass theorem by Ambrosetti-Rabinovitz since J' has no meaning. But under additional assumption (3.6) we shall prove a weak form of the Palais-Smale condition.

<u>Theorem 2</u> - Assume that the assumptions (3.2), (3.3), (3.4), (3.5), (3.6) hold.

Then any sequence $\{u_n\}$ such that

(3.7) $$|J(u_n)| \leq c_1$$

$$(3.8) \int_\Omega a(x,u_n,Du_n)Dv + \int_\Omega b(x,u_n,Du_n)v - \int |u_n|^{q-2}u_n v \to 0, \text{ for any } v \in W_0^{1,p}(\Omega) \cap L^\infty(\Omega)$$

has a converging subsequence.

<u>Proof</u> - the assumption $q<p$ and (3.7) imply that the sequence $\{u_n\}$ is bounded. For some subsequence $\{u_{n_j}\}$ weakly converging to some u we have

$$|u_{n_j}|^{q-2} u_{n_j} \to |u|^{q-2} u \quad \text{in} \quad L^{p'}(\Omega).$$

In order to prove the strong convergence of the sequence $\{u_{n_j}\}$ we shall use the assumption (3.6) and the results of the Section 2.4 of [BBM].

<u>Remark</u> - Critical points in the case $q>p$ are studied in a un published work by L. Boccardo and T. Gallouet when $N=1$.

References

[BBM] A. Bensoussan, L. Boccardo, F. Murat: On a nonlinear P.D.E. having natural growth terms and unbounded solutions; Ann. I.H.P. Anal. non. Lin. 5 (1988), 347-364.

[D] B. Dacorogna: Direct methods in the Calculus of Variations; Springer-Verlag, Berlin (1989).

[LU] O.A. Ladyzenskaja, M.M. Uralceva: Equations aux derivées partielles de type elliptique; Dunod, Paris (1968).

[S] G. Stampacchia: Le problème de Dirichlet pour les equations elliptiques du second ordre á coefficients discontinus; Ann. Inst. Fourier Grenoble 15 (1965), 189-258.

A MORSE THEORETIC APPROACH TO THE PRESCRIBING GAUSSIAN CURVATURE PROBLEM

KUNG CHING CHANG

Institute of Math. Peking University Beijing 100871 PRC

JIA QUAN LIU

Dept. of Math. Graduate School, Acad. Sinica, Beijing 100039 PRC

What function K can be the Gaussian curvature of a metric g on S^2, where $g = e^{2u}g_0, g_0$ is the standard metric and $u \in C^\infty(S^2)$? This is the Nirenberg's problem.

The answer of the question is equivalent to the solvability of the following equation:

$$-\triangle u(x) + 1 = K(x)e^{2u(x)} \qquad \forall x \in S^2. \tag{1}$$

According to the Gauss Bonnet's formula, a necessary condition: $\Lambda = \text{Max} K > 0$ is derived. What are the sufficient conditions?

By the variational approach, we consider the functional

$$J(u) = \int_{S^2}(|\nabla u|^2 + 2u) - \log \int_{S^2} Ke^{2u} \tag{2}$$

under the constraint

$$\int_{S^2} e^{2u} = 1 \tag{3}$$

where $\int$ denotes the average.

In his pioneer work, J.Moser[11] proved that $\Lambda > 0$ is also sufficient, if K is even. Geometrically, this gave an answer to the Nirenberg's problem for the real projective space. The proof is to find the minimum of J on the subspace:

$$\widehat{X} = \left\{ u \in H^1(S^2) | u(x) = u(-x) \text{ and } \int_{S^2} e^{2u} = 1 \right\}.$$

The coerciveness of J on $\widehat{X}$ follows from the inequality: $\forall \varepsilon > 0, \exists C_\varepsilon > 0$ such that for even u,

$$\int_{S^2} e^{2u} \leq C_\varepsilon \exp\left[\frac{1}{2-\varepsilon} \int_{S^2} |\nabla u|^2 + \int_{S^2} 2u \right]. \tag{4}$$

However, without the restriction on evenness, one has only

$$\int_{S^2} e^{2u} \leq \exp\left[\int_{S^2} (|\nabla u|^2 + 2u) \right] \tag{5}$$

The inequality (5) is sharp, see Hong[8] and Onofri[12].

According to (5), J is bounded from below. Unfortunately, the minimum of J on the manifold

$$X = \left\{ u \in H^1(S^2) | \int_{S^2} e^{2u} = 1 \right\}$$

is not achieved, if $K \not\equiv$ const. Indeed, for $K \equiv 1$,

$$J(u) = S(u) := \int_{S^2} (|\nabla u|^2 + 2u)$$

on X. $\forall \phi \in \text{Conf } (S^2)$, the conformal diffeomorphism on S^2, $\forall u \in C^\infty(S^2)$, let

$$u_\phi = u \circ \phi + \frac{1}{2} \log \det |\phi'|,$$

we have

$$S(u_\phi) = S(u),$$

and

$$\int_{S^2} e^{2u_\phi} = \int_{S^2} e^{2u}.$$

In particular, the set

$$M = \left\{ \frac{1}{2} \log \det |\phi'| | \phi \in \text{Conf } (S^2) \right\}$$

is the set of all minima of S on X.

$\forall (Q,t) \in S^2 \times [1,\infty)$, we introduce a conformal mapping

$$\phi_{Q,t} = \pi_Q^{-1} \circ m_t \circ \pi_Q,$$

where π_Q is the stereographic projection $S^2 \longrightarrow \hat{\mathcal{C}} = \mathcal{C} \cup \{\infty\}$, with Q as the north pole, and m_t is the dilation:

$$m_t z = t \cdot z \qquad \forall z \in \hat{\mathcal{C}}.$$

Let

$$\mathcal{A} = \{\phi_{Q,t} | (Q,t) \in S^2 \times [1,\infty)\}/S^2 \times \{1\},$$

be the subgroup of $\mathrm{Conf}(S^2)$, then the set

$$\left\{\frac{1}{2} \log \det |\phi'| \,\Big|\, \phi \in \mathcal{A}\right\} \cong \overset{\circ}{B}{}^3$$

coincides with the set M.

A direct computation shows that

$$e^{2\psi_{Q,t}} \longrightarrow 4\pi\delta(-Q) \quad \text{in} \quad \mathcal{D}'(S^2)$$

as $t \to +\infty$, where

$$\psi_{Q,t} = \frac{1}{2} \log \det |\phi'_{Q,t}|.$$

If $K(Q_0) = \Lambda$, then

$$J(u) \geq -\log \Lambda,$$

and

$$J(\psi_{-Q_0,t}) \to -\log \Lambda;$$

i.e., $\inf J = -\log \Lambda$. Therefore, if $K \not\equiv \mathrm{const}$, and if u_0 were the minimum of J, then u_0 would be the minimum of the functional

$$\hat{J}(u) = S(u) - \log \Lambda$$

on X. It follows that u_0 is of the form $\frac{1}{2}\log\det|\phi'|$ for some $\phi \in \mathcal{A}$; but this is impossible.

The above facts show that the Palais Smale condition for J fails in both cases either K is a constant or not.

In order to understand why and where the Palais Smale condition fails, the manifold X is reparametrized. Namely, we have

Theorem 1. $\forall u \in X$, there exists a unique $(Q,t) \in S^2 \times [1,\infty)/S^2 \times \{1\}$, such that

$$\int_{S^2} x_i e^{2u+\phi_{Q,t}} = 0, \quad i = 1,2,3.$$

Moreover, (Q,t) depends on u smoothly.

The proof is as follows: locally, one solves the system

$$\int_{S^2} x \circ \phi_{-Q,t} e^{2u} = \int_{S^2} x e^{2u\phi_{Q,t}} = 0$$

by IFT; and globally, we use an argument similar to that of the global implicit function theorem, see Chang Liu[5] for details.

Thus, the diffeomorphism

$$u \longmapsto (\omega, Q, t);$$

where $\omega = u_{\phi_{Q,t}}$, splits the manifold X into $X_0 \times \overset{\circ}{B}{}^3$, where

$$X_0 = \left\{ \omega \in X \mid \int_{S^2} x_i e^{2\omega} = 0, i = 1,2,3 \right\}.$$

There is a sharp Onofri's inequality due to A.S.Y. Chang and P. Yang[1]

$$\int_{S^2} |\nabla \omega|^2 \leq \frac{1}{1-a} S(\omega) \qquad (6)$$

for some $a > 0, \forall \omega \in X_0$.

Rewrite J in the form

$$J(u) = S(\omega) - \log \int_{S^2} K \circ \phi_{Q,t} e^{2\omega}.$$

Since $S(\omega)$ is coercive on X_0, in combining with the Sobolev's embedding theorem, one sees that a Palais Smale sequence $u_j = (\omega_j, Q_j, t_j)$ does not subconverge if and only if $t_j \to +\infty$. In other words, the lack of compactness of this variational problem is arisen from the openness of the ball $\overset{\circ}{B}{}^3$.

We compactify the manifold X by $X_0 \times \overline{B^3}$, and extend J into

$$\tilde{J}(u) = \begin{cases} J(u) & u = (\omega, Q, t) \in X_0 \times \overset{\circ}{B}{}^3, \\ S(\omega) - \log K(Q) & u = (\omega, Q) \in X_0 \times S^2 \end{cases}.$$

This is a continuous functional defined on a manifold with boundary $X_0 \times S^2$.

Let us recall the Morse theory under the general boundary condition, which is initiated by M. Morse and Van Schaack for domains in $\mathbb{R}^n$, and is extended by the authors[4] to Hilbert Riemannian manifolds with boundaries.

Assume that M is such a manifold modeled on a Hilbert space H, with the inner product (,). Assume that $\Sigma = \partial M$, the boundary of M is of codimension 1. A C^1 function f defined on $\overline{M}$ is said under the general boundary condition, if

(1) both f and $\hat{f} := f|_\Sigma$ possess only isolated critical points,
(2) f has no critical points on Σ.
Denote
$$\Sigma_- = \{x \in \Sigma | (n(x), f'(x)) \leq 0\}$$
where $n(x)$ is the outward unit normal vector at x.

Suppose that $a < b$ are regular values of f and $\hat{f}$, and that the $(PS)_c$ condition holds for both f and $\hat{f}$, $\forall c \in [a,b]$. The following augmented Morse inequalities hold:

$$\sum_{q=0}^{\infty}(m_q + \mu_q - \beta_q)t^q = (1+t)Q(t) \qquad (7)$$

where Q is a formal series with nonnegative coefficients,

$$m_q = q^{th} \quad \text{Morse type number of} \quad f \quad \text{in} \quad f^{-1}[a,b],$$

$$\mu_q = q^{th} \quad \text{Morse type number of} \quad \hat{f} \quad \text{in} \quad \hat{f}^{-1}[a,b],$$

and
$$\beta_q = \text{rank} \ H_q(f_b, f_a).$$

Noticing the following asymptotic expansion due to A.S.Y. Chang and P.Yang [2], see also Z.Han [7]:

$$\begin{aligned}J(u) = \ & S(\omega) - \log\{K(Q) + 2\triangle K(Q)t^{-1}\log t \\ & +O(|\nabla K(Q)|(t^{-1}\log^{\frac{1}{2}}tS(\omega)^{\frac{1}{2}})\} + O(t^{-2} + S(\omega)^2),\end{aligned} \qquad (8)$$

we are encouraged to count μ_q for $\tilde{J}$ by the use of (8). Unfortunately, $\tilde{J} \notin C^1(X_0 \times \overline{B}^3)$, because the differentiability is lost at the boundary $\Sigma = X_0 \times S^2$. The technical difficulty is overcome by the following

Theorem 2. If $\Lambda > 0$ and

$$\triangle K(x) \neq 0 \quad \text{whenever} \quad \nabla K(x) = 0 \quad \text{and} \quad K(x) > 0 \qquad (9)$$

then there exists a function $f \in C^1(X_0 \times \overline{B}^3)$ satisfying
(1) f possesses the same critical set as J,
(2) $\exists$ a neighbourhood U of Σ, if $K(Q) > 0$,

$$f(\omega, sQ) = S(\omega) - \log K(Q) - \frac{2\triangle K(Q)}{K(Q)}(1-s) \qquad (10)$$

$\forall (\omega, sQ) \in U$,

(3) f satisfies the (PS) condition.

The proof of this theorem depends on delicate asymptotic expansions and a careful construction, in combining with a fairly long computation, see Chang Liu[5].

Under the assumption (9) and the following assumption:

$$K \text{ possesses finite number of critical points,} \qquad (11)$$

one may count μ_q for f. According to Theorem 2, this is the same for $\tilde{J}$. Denote

$$\Omega = \{x \in S^2 | K(x) > 0, \triangle K(x) < 0\},$$

$$C_{r_0}(a,b) = \{x \in \Omega | K(x) \in (a,b), \quad \text{and} \quad x \text{ is a local maximum }\},$$

and

$$C_{r_1}(a,b) = \{x \in \Omega | K(x) \in (a,b), \quad \text{and} \quad x \text{ is a saddle point }\}.$$

If $0 \leq a < b$ are regular values of K, then

$$\mu_q = \begin{cases} 0 & q \geq 2, \\ \#C_{r_q}(a,b) & q = 0,1, \end{cases} \qquad (12)$$

for $M = f^{-1}[-\log b, -\log a]$.

Since the function f satisfies the general boundary condition if J has only isolated critical points, the Morse inequalities (7) with such μ_q improve a series of results known in literatures.

First, let $p = \#C_{r_0}(0,+\infty), q = \#C_{r_1}(0,+\infty)$. Since $f^{-1}(-\infty,+\infty)$ is contractible, we have

$$\beta_q = \delta_{q0}.$$

Thus, (1) is solvable if

$$p \neq q + 1. \qquad (13)$$

This improves a result due to A.S.Y.Chang and P.Yang[2], in which K is positive and nondegenerate.

Second, we turn out to improve a result due to W.X.Chen, W.Y.Ding[6] and C.Hong[9]:

Suppose that there exists two points $P_0, P_1 \in S^2$, where P_0 is a local maximum of K with

$$m := K(P_0) \leq K(P_1).$$

Let Γ be the set of pathes connecting P_0 and P_1. Assume that

$$\nu = \sup_{h \in \Gamma} \inf_{x \in h} K(x) < m$$

and that
$$C_{r_1}(\nu - \varepsilon_1, m) = \emptyset \tag{14}$$
for some $\varepsilon_1 > 0$. Then (1) is solvable.

In Chen Ding[6], P_0 and P_1 are assumed to be maxima. In both Chen Ding [6] and Hong [9], besides (14), they also assumed $C_{r_0}(\nu - \varepsilon_1, m) = \emptyset$.

Third, some other results may be derived from the linking argument.

(i) Suppose that there exists an isolated saddle point $Q_0 \in \Omega$ and a C^1 curve $\gamma : S^1 \to S^2$ passing through Q_0, at which γ' lies along the direction of the eigenvector associate with the larger eigenvalue of $\nabla^2 K(Q_0)$. If
$$\min_{x \in \gamma} K(x) = K(Q_0),$$
then (1) is solvable.

(ii) Suppose that there exists a Jordan curve $\gamma : S^1 \to S^2$, such that $K(x) = \Lambda, \forall x \in \gamma$, and that $S^2 \setminus \gamma(S^1)$ contains two components each including a point below Λ. Then (1) is solvable.

The last one improves a theorem due to C.Hong and Y.D.Wang[10], in which the condition that there is no critical point of K above the level
$$\nu = \sup_{Q \in \Gamma} \inf_{x \in Q} K(x)$$
where Γ is the set of all topological 2 disks on S^2 with boundary γ, was assumed. But this condition is superfluous. One of the advantages of the Morse theoretic approach is that one may estimate the critical groups for critical points obtained by links Cf.Chang[3].

Finally, even the condition (11) can be dropped. Namely, we have the following

Theorem 3. Assume (9) and
$$\deg(\nabla K, \Omega, \theta) \neq 1.$$
The problem (1) is solvable.

One uses the perturbation argument, according to the Sard's Theorem, this is reduced to the Morse inequalities.

References

[1] A.S.Y. Chang, P.Yang; JDG 23 (1988), 259–296.
[2] ——; Acta Mathematica, 159 (1987), 214–259.

[3] K.C.Chang; Critical groups, Morse theory and applications to semilinear elliptic BVP; Chinese Math. towards 21 Century, (1991) 41–65.
[4] K.C.Chang, J.Q.Liu; J.System Sci. and Math. Sci. 4 (1991), 78–83.
[5] ——; On the Nirenberg's Problem; Preprint, Inst. of Math. Peking Univ. (1992).
[6] W.X.Cheng, W.Y.Ding; TAMS 303 (1987), 365–382.
[7] Z.C.Han; Duke Math. J. 61 (1990), 679–703.
[8] C.W.Hong; PAMS, 97 (1986), 737–747.
[9] ——; PDE (Chinese), 1 (1987), 13–20.
[10] C.W.Hong, Y.D.Wang; Prescribing Gaussian curvature and scalar curvature, preprint.
[11] J.Moser; On a nonlinear problem in differential geometry, Dynamical Systems, Acad. Press (1973).
[12] E. Onofri; Comm. Math. Phys. 86 (1982), 321–326.

EVOLUTION PROBLEM OF YANG–MILLS FLOW OVER 4–DIMENSIONAL MANIFOLD

CHEN YUN-MEI* AND SHEN CHUN-LI**

*Department of Mathematics, University of Florida
Gainesvill, FL 32611, USA
**Department of Mathematics, East China Normal University
Shanghai 200062, P.R.China

§1. Introduction

Let $P(M^n, G)$ be a principal bundle over n-dimensional compact Riemannian manifold M^n ($n \geq 4$) with compact, simple Lie structure group G. Let A_t be a Yang-Mills flow with an initial connection A_0, i.e.,

$$\begin{cases} \dfrac{\partial A_t}{\partial t} = -2 D^*_{A_t} F_{A_t}, \\ A_t|_{t=0} = A_0 \end{cases}$$

where F_{A_t} are the curvatures of A_t, D_{A_t} the gauge-covariant differentiation, and $D^*_{A_t}$ the adjoint operator of D_{A_t} with respect to the Riemannian metric of M and Cartan-Killing inner product of G. It is well known that the solution of Yang-Mills flow exists for short time. But the problem of long time existence for Yang-Mills flow is rather complicated. In general, Yang-Mills flow may blow up at some finite time. In this paper, we assume that the Yang-Mills flow A_t is a regular global flow for any time t, and our interest is only to discuss the limiting situation of A_t when $t \to \infty$. The main result is the following.

Main Theorem *Let $P(M^4, G)$ be a principal bundle over a 4-dimensional compact Riemannian manifold (M^4, g) with compact, simple Lie structure group G, $k(P)$ Pontrjagin index of P. Let A_t be a regular global Yang-Mills flow with an initial connection A_0 for any time $t \geq 0$. Then we can take a subsequence t_i suitably, and find a sequence of gauge transformations σ_i, a subset $\Sigma_\infty = \{p_1, \cdots, p_l\}$ of finite points in M, such that $\{\sigma_i^*(A_{t_i})\}$ C^∞-converges to a regular Yang-Mills connection A_∞ on $M \setminus \Sigma_\infty$, A_∞ can be extended to a regular Yang-Mills connection $\tilde{A}_\infty$*

---

† The first author was partially supported by NSF DMS-9123532. The second author was partially supported by National Science Foundation of China.

over whole M, but $\tilde{A}_\infty$ may belong to an another bundle Q which is an extension of $P|_{M\setminus\Sigma_\infty}$. Furthermore, for each singular point p_α, $(\alpha = 1, \cdots, l)$, there corresponds a principal G-bundle P_α over (S^4, can) with Pontrjagin index k_α (where 'can' means the standard metric on S^4), and a non-flat Yang-Mills connection A_α on P_α, such that

$$k(Q) + \sum_{\alpha=1}^{l} k(P_\alpha) = k(P), \quad \text{and} \quad YM(\tilde{A}_\infty) + \sum_{\alpha=1}^{l} YM(A_\alpha) = YM_\infty,$$

where $YM(A)$ is the Yang-Mills action of connection A, and $YM_\infty = \lim_{t\to\infty} YM(A_t)$.

§2. Energy inequality, Bochner type inequality, and ε-regularity theorem for Yang-Mills flow

In [4] we have proved the following results:

Energy Inequality

$$\int_0^t \int_M |J_{A_t}|^2 \sqrt{g} dx dt + YM_t = YM_0 ,$$

where $J_{A_t} = 2 D^*_{A_t} F_{A_t}$, $YM_t = YM(A_t)$. In particular, $YM_t \leq YM_0$.

Bochner Type Estimate

$$(\frac{\partial}{\partial t} - \Delta)|F_{A_t}| \leq C(|F_{A_t}| + |Rm|)|F_{A_t}| ,$$

where $|Rm|$ is the norm of Riemannian curvatures of metric g, Δ is the Laplacian of metric g and constant C dependents only on the geometry of M.

For any $(x_0, t_0) \in M \times R_+$, where R_+ is the set of all nonnegative real numbers, let

$$T_R(x_0, t_0) = \{(x,t) \in M \times R_+ | -4R^2 < t - t_0 < -R^2\} ,$$

$$G_{(x_0,t_0)}(x,t) = (4\pi(t_0-t))^{-n/2} \exp\left\{-\frac{|x_0-x|^2}{4(t_0-t)}\right\}, \quad t < t_0,$$

where $|x_0 - x|$ means the Riemannian distance between points x_0 and x, and R_M a lower bound for the injectivity radius of M. For $0 < R < R_M$, we define the function $\Psi_{(x_0,t_0)}(R)$ for Yang-Mills flow A_t as follows:

$$\Psi_{(x_0,t_0)}(R) = R^2 \int_{T_R(x_0,t_0)} |F|^2 G_{(x_0,t_0)} \phi^2 \sqrt{g} dx dt ,$$

where $\phi \in C_0^\infty(B_{R_M}(x_0))$ is a suitable cut-off function satisfying $0 \leq \phi \leq 1$, $\phi \equiv 1$ on $B_{R_M/2}(x_0)$ and $|\nabla \phi| \leq C/R_M$.

ε-Regularity Theorem *Suppose $A(t)$ is a regular solution of Yang-Mills flow, $t \in [-T, T]$, and $T \leq R_M^2$. There exists a sufficiently small constant $0 < \varepsilon_0 < R_M$, depending only on M and the initial Yang-Mills action YM_0, such that if for some $0 < R < \min(\varepsilon_0, \sqrt{T}/2)$ and $x_0 \in M$, the inequality $\Psi_{(x_0,0)}(R) \leq \varepsilon_0$ is satisfied, then there holds*

$$\sup_{P_{\delta R}(x_0,0)} |F(A(t))|^2 \leq \frac{c}{(\delta R)^4}$$

with constant c depending only on M, YM_0, ε_0, constant $0 < \delta < 1/4$ depending only on M, YM_0, R, and $P_{\delta R}(x_0, 0) = \{(x, t) \in M \times R \mid |x - x_0| < \delta R, |t| < (\delta R)^2\}$. In particular, when $\dim M = n = 4$, then δ is independent on R.

§3. The sketch of the proof of main theorem

From energy inequality, we can choose a sequence $\{t_m\}$ satisfying the following two conditions as $m \to \infty$:

$$\int_M |J_{A_{t_m}}|^2 \sqrt{g} dx \longrightarrow 0 \quad \text{and} \quad \int_{t_m-1}^{t_m} \int_M |J_{A_t}|^2 \sqrt{g} dx dt \longrightarrow 0.$$

1° Using Sedlacek's technique [5,Proposition 3.3], we can divide the whole manifold M into two parts : the 'good' part $M \setminus \Sigma_\infty$ and the 'bad' part Σ_∞, according to the properties of function Ψ. We can choose a countable many covering $\{B_{R_i}(x_i)\}$ of $M \setminus \Sigma_\infty$, such that $\Psi_{(x_i,t_m)}(R_i/\delta) \leq \varepsilon_0$ holds for sequence $\{t_m\}$, (or its some subsequence, if necessary). Using the similar argument as [1,p.100], we can prove that Σ_∞ is a finite set of points in M, i.e., $\Sigma_\infty = \{p_1, \cdots, p_l\}$.

2° Using ε-regularity theorem for every ball $B_{R_i}(x_i)$ in the above covering of $M \setminus \Sigma_\infty$, there holds $|F_{A_t}| < c/R_i^2$ on $P_{R_i}(x_i, t_m)$ for any t_m. Namely, the C^0-norm of F_{A_t} are bounded on a rather small domain in $P_{R_i}(x_i, t_m)$ uniformly (with respect to t_m). In particular, for the sequence of connections $\{A_{t_m}\}$, the norms $|F_{A_{t_m}}|$ are bounded uniformly on B', where B' is a rather small ball in B.

3° We fix any one of the balls $B_{R_i}(x_i)$ in this covering, and denote it simply by $B = B_{R_i}(x_i)$. Using the Bochner type inequality, not only $|F_{A_t}|$ but also the norms of higher gauge-covariant derivatives $|\nabla_{A_t}^k F_{A_t}|$ are uniformly bounded for all $k = 1, 2, \cdots$, on a rather small domain P' in $P = P_{R_i}(x_i, t_m)$, i.e., $|\nabla_{A_t}^k F_{A_t}| < C_k$, for any k on P'. In particular, for the sequence of connections $\{A_{t_m}\}$, the norms $|\nabla_{A_{t_m}}^k F_{A_{t_m}}|$ are bounded uniformly on B', where B' is a rather small ball in B.

4° Now we have controlled the curvatures $F_{A_{t_m}}$ and their higher gauge-covariant derivatives. But it is not enough to get the C^∞-convergence for a sequence of connections, and we have to esimate the norms of connections and their higher derivatives. To this end, we use the Uhlenbeck's Coulomb gauge fixing arguments. We want to estimate the norms of the connections and their higher derivatives under the Coulomb gauge.

Using bootstrapping technique, under the Coulomb gauge σ_m, the Sobolev norms of connections $\tilde{A}_m = \sigma_m^*(A_{t_m})$ can be estimated by

$$\|\tilde{A}_m\|_{L^2_{l+1}(B')} \leq C_l(\|F_{A_{t_m}}\|_{C^0(B)} + \sum_{k=1}^{l} \|\nabla^k_{A_{t_m}} F_{A_{t_m}}\|_{L^2(B)}),$$

where B' is a rather small domain in B ([2, Lemma 2.3.11]). As the right side of this estimate is uniformly bounded, $\|\tilde{A}_m\|_{L^2_{l+1}(B')}$ is also uniformly bounded for every l. Then, from Sobolev imbedding theorem, the C^l-norms of $\{\tilde{A}_m\}$ are uniformly bounded for $l = 0, 1, 2, \cdots$. Using the Arzela-Ascoli theorem, we can choose a subsequence of $\{t_m\}$ such that $\{\tilde{A}_m\}$ C^∞-converges to a limiting connection A_∞ over B'. Since

$$\int_M |J_{A_{t_m}}|^2 \sqrt{g} dx \longrightarrow 0$$

as $m \to \infty$, so $J_{A_\infty} = 0$. Hence the limiting connection A_∞ is a Yang-Mills connection over B'.

5° Using Uhlenbeck's patching arguments [3], these individual limiting connections A_∞ on each ball can be glued together, and become a limiting connection over $M \setminus \Sigma_\infty$.

6° Up to now, we could choose a sequence $\{t_i\}$ such that, under suitable gauge transformations σ_i, $\sigma_i^*(A_{t_i})$ C^∞-converges to a limiting connection A_∞ on $M \setminus \{p_1, \cdots, p_l\}$. Using Uhlenbeck's removable singularity theorem, these singular points p_α could be removed. Obviously, the sequence $\{\sigma_i^*(A_{t_i})\}$ is a 'good' sequence in the sense of Taubes [6], so the proof of main theorem is completed from [6].

References

[1]. Y.M.Chen and M.Struwe, *Existence and partial regularity for heat flow for harmonic maps*, Math. Zeit. **201** (1989), 83–103.
[2]. S.K.Donaldson and P.B.Kronheimer, *The geometry of four-manifolds*, Clarendon Press, Oxford, 1990.
[3]. D.S.Freed and K.K.Uhlenbeck, *Instantons and four-manifolds*, Springer-Verlag, New York, 1984.
[4]. C.L.Shen, *Monotonicity formula and partial regularity results in Yang-Mills flow*, Qualitative aspects and applications of nonlinear evolution equations (T.T.Li and P. de Mottoni, ed.), World Scientific, Singapore, 1991, p. 248-251.
[5]. S.Sedlacek, *A direct method for minimizing the Yang-Mills functional over 4-manifolds*, Commun. Math. Phys. **86** (1982), 515–527.
[6]. C.H.Taubes, *Path-connected Yang-Mills moduli spaces*, Jour. Diff. Geom. **19** (1984), 337–392.

PERIODIC SOLUTIONS FOR A CLASS OF SINGULAR NONAUTONOMOUS SECOND ORDER SYSTEM IN A POTENTIAL WELL

V. COTI ZELATI(*), LI SHUJIE(**) and WU SHAOPING(***)

(*) Ist. Mat., Fac. Architettura, 80134 Napoli, Italy
(**) Academia Sinica, Institute of Math., 100080 Beijing, China
(***) Zhejiang Univ. Dept. of Math., 310027 Hangzhou, China

0. INTRODUCTION.

In this paper we study the existence of periodic solutions for dynamical systems of the form

(HS) $$-\ddot{u} + Au - \nabla V(t, u) = 0$$

where V is singular in the sense that $V \in C^1(\mathbb{R} \times \Omega; \mathbb{R})$, with Ω open subset of $\mathbb{R}^N$ and $V(t, x) \to +\infty$ as $x \to \partial\Omega$. This problem has been recently studied, using variational methods, by many authors. We recall here, [AC1], [BR], [DG], [G] and [R] papers which deal with the case $\Omega = \mathbb{R}^N \setminus \{0\}$.

Here we study the case in which Ω is a bounded open set of $\mathbb{R}^N$. Such a problem has been studied in [B], [AC2] (see also [GR] for the case Ω unbounded). Here we give a simpler proof of the results contained there, in a slightly different situation.

Our result is the following

Theorem. *Suppose V satisfies*

(V_0) $V \in C^1(\mathbb{R} \times \Omega; \mathbb{R})$ *is T periodic in t, while $\Omega \subset \mathbb{R}^N$ is a bounded open set such that $0 \in \Omega$ and $\partial\Omega$ is of class C^1;*

(V_1) $V(t, y) = 0(|y|^2)$ *at $y = 0$;*

(V_2) There are constants $\alpha > 0$ and $\theta \in (0, \frac{1}{2})$ such that
$$V(t, x) \leq \theta(\nabla V(t, x), x) \quad \forall x \in (\partial\Omega)_\alpha$$
where $(\partial\Omega)_\alpha = \{x \in \Omega \mid \text{dist}(x, \partial\Omega) \leq \alpha\}$;

(V_3) $\dfrac{\partial V}{\partial t} \leq \delta V + \mu$ where δ, μ are positive constants;

(V_4) $\liminf_{\varepsilon \to 0} \varepsilon M_\varepsilon > 0$, where $M_\varepsilon(t) = \inf\{V(t, x) \mid x \in \Omega \cap (\partial\Omega)_\varepsilon\}$, uniformly for $t \in [0, T]$

Then

(a) If 0 is not an eigenvalue of $L = -\dfrac{d^2}{dt^2} + A$ (with periodic boundary conditions) (HS) has a non trivial T-periodic solution.

(b) If 0 is an eigenvalue of L and either

(V_6') $\exists r > 0$ such that $V(t, x) \geq 0 \; \forall 0 < |x| \leq r \; \forall t \in [0, T]$ or

(V_6'') $\exists r > 0$ such that $V(t, x) < 0 \; \forall 0 < |x| \leq r \; \forall t \in [0, T]$

then (HS) has a non trivial T-periodic solution.

With respect to the results contained in [B], here we allow V to depend on time and we do not require that $V(t, x) \geq V(t, 0) \equiv 0 \; \forall (t, x)$. Remark that, being V time dependent, one cannot just replace 0 by the infimum of V over Ω. Always with respect to [B] we also give different conditions which V has to satisfy close to $\partial\Omega$, and our proof is simpler that the one given in [B]. On the other hand, our assumption (V_2) implies that Ω is star-shaped.

With respect to the results of [AC2], we do not require Ω to be convex and V to be positive everywhere; but we cannot prove minimality of the period for the solution we find (for the time independent case).

The paper is organized in two sections: in section 1 we prove the Palais-Smale condition and some technical lemmas. The proof of the Palais-Smale condition is based on the conservation of mechanical energy. In section 2 existence of a critical point for the functional f associated to (HS) is proved, using a finite-dimensional approximation, together with a linking argument.

1. PRELIMINARY LEMMAS.

Let E be a Hilbert space with orthonormal basis $\{e_1, e_2, \dots\}$, let $E_n = \text{span}\{e_1, \dots, e_n\}$, Λ be an open set in E, and $f : \Lambda \to \mathbb{R}$ be a C^1 function. Set $\Lambda_n = E_n \cap \Lambda$ and $f_n \doteq f|_{\Lambda_n}$.

We first recall the following well-known definitions.

If any sequence $x_n \in \Lambda$, satisfying $|f(x_n)| \leq M < +\infty$ and $\nabla f(x_n) \to 0$, possesses a convergent subsequence in Λ then we say that f satisfies the (PS) condition.

If any sequence $x_n \in \Lambda_n$ satisfying $f_n(x_n) \leq M < +\infty$ and $\nabla f_n(x_n) \to 0$ possesses a convergent subsequence in Λ then we say that f satisfies the $(PS)^*$ condition in Λ.

Throughout this paper we take $E = W_T^{1,2} = H^1(S_T^1, \mathbb{R}^N)$ and, given an open set $\Omega \subset \mathbb{R}^N$, we set $\Lambda = \{x(t) \in E \mid x(t) \in \Omega \ \forall t \in [0,T]\}$. Let $\{e_1, \ldots, e_n, \ldots\}$ be a smooth orthonormal basis for H^1 and

$$f(x) = \int_0^T \frac{1}{2}|\dot{x}(t)|^2 + \int_0^T \frac{1}{2}(Ax(t), x(t)) - \int_0^T V(t, x(t))$$

It is well known that $f \in C^1(\Lambda, \mathbb{R})$ and that x is a T-periodic solution of (HS) in H^1 if and only if it is a critical points of $f(x)$. The following two lemmas play an important role in proving our theorem.

Lemma 1.1. *Suppose V satisfies (V_0), (V_2), (V_3) and (V_4). Then f satisfies (PS) and $(PS)^*$ conditions in Λ.*

Proof. We will only show that (PS) holds. Consider $\{u_n\} \subset \Lambda$ such that

$$f(u_n) \le c, \qquad f'(u_n) \to 0$$

Since

$$f(u_n) = \frac{1}{2}\int_0^T |\dot{u}_n|^2 + \frac{1}{2}\int_0^T (Au_n, u_n) - \int_0^T V(t, u_n),$$

$$(f'(u_n), u_n) = \int_0^T |\dot{u}_n|^2 + \int_0^T (Au_n, u_n) - \int_0^T (\nabla V(t, u_n), u_n)$$

and since from (V_2) it follows

$$V(t, y) \le \theta(\nabla V(t, y), y) + c \qquad \forall (t, y) \in S^1 \times \Omega$$

we have that

$$\frac{1}{2}\int_0^T |\dot{u}_n|^2 + \frac{1}{2}\int_0^T (Au_n, u_n) - f(u_n)$$

$$\le \theta[\int_0^T |\dot{u}_n|^2 + \int_0^T (Au_n, u_n) - (f'(u_n), u_n)] + cT,$$

namely

$$(\frac{1}{2} - \theta)\int_0^T |\dot{u}_n|^2 \le f(u_n) - \theta(f'(u_n), u_n) - (\frac{1}{2} - \theta)\int_0^T (Au_n, u_n) + cT,$$

then

$$(\frac{1}{2} - \theta)\left[\int_0^T |\dot{u}_n|^2 + |u_n|^2\right]$$

$$\le f(u_n) + \theta\|f'(u_n)\|\|u_n\| - (\frac{1}{2} - \theta)\left[\int_0^T (Au_n, u_n) - \int_0^T u_n^2\right] + cT.$$

and we deduce

$$(\frac{1}{2} - \theta)\|u_n\|^2 \le f(u_n) + \theta\|f'(u_n)\|\|u_n\| + c(\frac{1}{2} - \theta)\|u_n\|_2^2 + cT.$$

Since Ω is bounded, $\|u_n\|_2 \le const$ so that

$$\|u_n\|^2 \le c + c\varepsilon_n\|u_n\|,$$

with $\varepsilon_n \to 0$. Hence $\|u_n\| \le const$, which implies $u_n \to u^*$ weakly in H^1 and uniformly so that $u_n(t) \to u^*(t) \in \bar\Omega$ uniformly.

We claim that there exists $t_0 \in S^1$ such that $u^*(t_0) \in \Omega$. In fact, otherwise $u^* \in \partial\Omega$, $\forall t \in [0,T]$ and from the uniform convergence we deduce that for any $\varepsilon > 0$ there exists k such that $\forall n \ge k$ we have

$$\text{dist}(u_n(t), \partial\Omega) \le \varepsilon \qquad \forall t \in [0,T]$$

Using (V_4) we get

$$V(t, u_n(t)) \to \infty \quad \text{uniformly} \quad \forall t \in [0,T]$$

and

$$\int_0^T V(t, u_n) \to +\infty, \quad \text{as} \quad n \to \infty$$

contradiction with $|f(u_n)| \le c$.

Let

$$t_1 = \inf\{t < t_0 \text{ such that } u^*(s) \in \Omega, \forall s \in (t, t_0]\}$$

and

$$t_2 = \sup\{t > t_0 \text{ such that } u^*(s) \in \Omega, \forall s \in [t_0, t)\}$$

We have $t_1 < t_0 < t_2$, since u^* is continuous. Taking $\varphi \in C_0^\infty((t_1, t_2), \mathbb{R}^N)$, we have

$$\int_0^T \dot u_n \dot\varphi + \int_0^T (Au_n, \varphi) - \int_0^T (\nabla V(t, u_n), \varphi) = < f'(u_n), \varphi >.$$

Since

$$\int_0^T (Au_n, \varphi) \to \int_0^T (Au^*, \varphi)$$

$$\int_0^T (\nabla V(t, u_n), \varphi) \to \int_0^T (\nabla V(t, u^*), \varphi)$$

and
$$< f'(u_n), \varphi > \to 0,$$
we have, passing to the limit,
$$\int_0^T \dot{u}^* \dot{\varphi} + \int_0^T (Au^*, \varphi) - \int_0^T (\nabla V(t, u^*), \varphi) = 0, \ \forall \varphi \in C_0^\infty((t_1, t_2), \mathbb{R}^N).$$
So u^* is a classic solution of
$$-\ddot{u}^* + Au^* = \nabla V(t, u^*) \ \forall t \in (t_1, t_2)$$
It is easy to see that
$$\frac{d}{dt}(\frac{1}{2}|\dot{u}^*|^2 + \frac{1}{2}(Au^*, u^*) + V(t, u^*)) = \frac{\partial V}{\partial t}(t, u^*), \ \forall t \in (t_1, t_2).$$
Setting $E_t = \frac{1}{2}|\dot{u}^*(t)|^2 + \frac{1}{2}(Au^*(t), u^*(t)) + V(t, u^*(t))$, we get, by integrating,
$$\begin{aligned}E_t - E_0 =& \frac{1}{2}|\dot{u}^*(t)|^2 + \frac{1}{2}(Au^*(t), u^*(t)) + V(t, u^*(t)) \\ & - \frac{1}{2}|\dot{u}^*(t_0)|^2 - \frac{1}{2}(Au^*(t_0), u^*(t_0)) - V(t_0, u^*(t_0)) \\ =& \int_{t_0}^t \frac{\partial V}{\partial t}(t, u^*)\end{aligned}$$
and
$$V(t, u^*(t)) = E_0 - \frac{1}{2}|\dot{u}^*(t)|^2 - \frac{1}{2}(Au^*(t), u^*(t)) + \int_{t_0}^t \frac{\partial V}{\partial t}(t, u^*(t))$$
Recalling $|u^*(t)| \leq const.$ and (V_3), we get
$$V(t, u^*(t)) \leq c + \int_{t_0}^t \frac{\partial V}{\partial t}(t, u^*(t)) \leq c + \delta \int_{t_0}^t V(t, u^*(t))$$
Using Gronwall Lemma, we finally get
$$V(t, u^*(t)) \leq c_1 \exp(c_2 t), \ \forall t \in [t_0, t_2)$$
which gives $u^*(t) \in \Omega \ \forall t \in [t_0, t_2]$. The same argument holds for $t \in [t_1, t_0]$. So $u^*(t) \in \Omega$, $\forall t \in [0, T]$, namely, $u^* \in \Lambda$. By a standard procedure, one then shows that $u_n \to u^*$ in H^1 and $f'(u^*) = 0$. This completes the proof. □

We now study the behavior of f near the boundary of Λ_n.

Lemma 1.2. *Suppose* (V_0), (V_2), (V_3) *and* (V_4) *hold. Then for any fixed* k *there exists an* $\varepsilon > 0$ *such that*

$$(a) \quad f_k(u_n) \to -\infty \quad \text{if and only if} \quad u_n \to \partial \Lambda_k$$
$$(b) \quad \frac{d}{d\lambda} f_k(\lambda u) < 0 \quad \text{as} \quad \lambda u \in (\partial \Lambda_k)_\varepsilon$$

where $(\partial \Lambda_k)_\varepsilon = \{v \in \Lambda_k \mid \text{dist}(v, (\partial \Lambda_k)) < \varepsilon\}$.

Proof.
(a) The necessary condition is obvious and we only show that the condition is sufficient. Since Λ_k is finite dimensional and bounded, we can assume

$$u_n \to \bar{u} \in \partial \Lambda_k.$$

Since the basis $\{e_1 \ldots, e_k\}$ is made of smooth function, $\bar{u}$ is smooth too. Let

$$d(t) = \text{dist}(\partial \Omega, \bar{u}(t))$$

Since $\bar{u} \in \partial \Lambda_k$, without loss of generality, we can assume that $d(0) = 0$. Since $\partial \Omega$ is of class C^1 we get that $d(t)$ is C^1 in a neighborhood of $t = 0$. Since $d(t) \geq 0$ and $d(0) = 0$ we have that $d'(0) = 0$ so that for any $\delta > 0$, $\exists \mu$ such that $\forall t \in (-\mu, \mu)$ we have $d(t) \leq \delta t \leq \delta \mu$. Hence for n large enough, we have

$$V(t, u_n(t)) \geq M_{2\delta\mu}, \quad \forall t \in (-\mu, \mu)$$

and from (V_4) we deduce that, for δ small,

$$\int_{-\mu}^{\mu} V(t, u_n(t)) \geq 2\mu M_{2\delta\mu} = \frac{1}{\delta} 2\delta\mu M_{2\delta\mu} \geq \frac{\sigma}{\delta}$$

Therefore

$$\int_{-\mu}^{\mu} V(t, u_n(t)) \geq \frac{\sigma}{\delta} \to +\infty$$

as $\delta \to 0$.

To prove (b), let us remark that

$$\frac{d}{d\lambda} f_k(\lambda u) = \frac{\lambda}{2} \int_0^T |\dot{u}|^2 - \int_0^T (\nabla V(t, \lambda u), u)$$
$$\leq \frac{\lambda}{2} \int_0^T |\dot{u}|^2 - \frac{1}{\theta\lambda} \int_0^T V(t, \lambda u)u$$

and the result follows as in (a). □

Remark 1.3. Since $\frac{d}{d\lambda} f_k(\lambda u) = (f'_k(\lambda u), u)$ lemma 1.3 implies the set $\{f_k \geq -M\}$ is star shaped as M is large enough.

2. THE PROOF OF THE THEOREM..

Let us start by proving an abstract result from which Theorem will follow. As in section 1 we take E to be a Hilbert space with basis $(e_1, \ldots, e_n, \ldots)$ and let $E_n = \text{span } \{e_1, \ldots, e_n\}$. Λ will denote an open set in E such that $0 \in \Lambda$ and $\Lambda_n = \Lambda \cap E_n$. Given $f : \Lambda \to \mathbb{R}$, we set

$$f_n = f|_{\Lambda_n}$$
$$f_{n,M} = \{u \in \Lambda_n \mid f_n(u) \geq M\}$$
$$f_n^M = \{u \in \Lambda_n \mid f_n(u) \leq M\}$$

Following [L], we give the following definition:

Definition 2.1. Let $f : \Lambda \to \mathbb{R}$ be a C^1 function where Λ is an open set of E with $0 \in \Lambda$ and $B(0,r) \doteq \{x \in \Lambda \mid \|x\| \leq r\} \subset \Lambda$. Suppose that there are two positive constants b and r such that

$$\begin{array}{ll} f(x) \geq 0 & \forall x \in B_1 \doteq B(0,r) \cap E^1, \\ f(x) \geq b & \forall x \in \partial B_1, \\ f(x) \leq 0 & \forall x \in B_2 \doteq B(0,r) \cap E^2 \end{array}$$

where $E = E^1 \oplus E^2$ with $\dim E^2 = k < \infty$. Then we say that f has an k-dimensional local linking at the origin.

It is easy to see that if f has a local linking at the origin, then 0 is a trivial critical point of f.

Lemma 2.2. Let $f \in C^1(\Lambda; \mathbb{R})$ be such that

(f_1) f verifies (PS) and $(PS)^*$ conditions in Λ
(f_2) f has a n-dimensional local linking at the origin
(f_3) $\forall x \in \Lambda \ \mathbb{R}_+ x \cap \partial \Lambda \neq \emptyset$
(f_4) $\forall n \in \mathbb{N} \ f_n(x) \to -\infty$ implies $x \to \partial \Lambda_n$
(f_5) $\forall n \in \mathbb{N} \ \exists \varepsilon_n > 0$ such that $[f_n(\lambda x)]'_\lambda < 0$ as $\lambda x \in (\partial \Lambda_n)_{\varepsilon_n} = \{y \in \Lambda_n \mid \text{dist}(y, \partial \Lambda_n) < \varepsilon_n\}$

then f has a non trivial critical point in Λ.

Proof. First of all, from (f_4) and (f_5), one easily deduces that

$$\forall n \in \mathbb{N} \quad \exists M_n \quad \text{such that} \quad [f'_n(\lambda u)]'_\lambda < 0 \quad \forall \lambda u \in f_n^{-M}$$

which implies that the set $f_{n,-M}$ is star-shaped (in particular it is diffeomorfic to the closed ball in Λ_n).

Suppose 0 is the only critical point for f in Λ. The $(PS)^*$ condition implies that there exists $n_0 \geq k$ (so that $E_{n_0} \supset E_k$) such that f_{n_0} has no critical point in the set

$$(\Lambda_{n_0} \setminus B(0,r)) \cap f_{n_0}^0$$

Take $M = \max\{M_{n_0}, M_{n_0+1}\}$. Since f_{n_0} satisfies the (PS) condition in Λ_{n_0}, we can construct a negative pseudo gradient flow $\eta: [0,1] \times \Lambda_{n_0} \to \Lambda_{n_0}$ such that

$$\begin{array}{lll} (a) & \eta(0,x) = x & \forall x \in \Lambda_{n_0} \\ (b) & f(\eta(1,x)) = -M & \forall x \in \partial B_2 \end{array}$$

Here B_2 denotes the ball of radius r in E^k.

Take $e_{n_0+1} \in (E_{n_0+1} \setminus E_{n_0}) \setminus \{0\}$ and $R > 0$ such that $f_{n_0+1}(Re_{n_0+1}) = -M$ (which exists by (f_3)).

Let

$$\psi(t,y) = (1-t)y + tRe_{n_0+1} \quad \forall y \in \eta(1, \partial B_2) \quad \forall t \in [0,1]$$

Then $\psi(t,y) \neq 0$ $\forall t \in [0,1]$ and $\forall y \in \eta(1, \partial B_2)$, since $y \in E_{n_0} \setminus \{0\}$, $e_{n_0+1} \in (E_{n_0+1} \setminus E_{n_0}) \setminus \{0\}$.

$\forall z \in \psi([0,1] \times \eta(1, \partial B_2))$, there exists $\lambda > 0$, such that $f_{n_0+1}(\lambda z) = -M$, which implies $[f_{n_0+1}(\lambda z)]'_\lambda < 0$. The Implicit Function Theorem then implies that there exists a continuous function $\lambda: \psi([0,1] \times \eta(1, \partial B_2)) \to \mathbb{R}$ such that $f_{n_0+1}(\lambda(z)z) = -M$. Such a λ satisfies $\lambda(z) = 1$ for $z \in \eta(1, \partial B_2) \cup (Re_{n_0+1})$.

Setting $\eta(x_2) = \eta(1, x_2)$, we define

$$\Phi(t, x_2) = \begin{cases} x_2, & x_2 \in B_2, t = 0, \\ \eta(2t, x_2), & x_2 \in \partial B_2, 0 \leq t \leq \frac{1}{2} \\ \lambda(\Psi(2t-1, \eta(x_2))\Psi(2t-1), \eta(x_2)), & x_2 \in \partial B_2, \frac{1}{2} \leq t < 1 \\ re_{n_0+1} & x_2 \in B_2, t = 1. \end{cases}$$

Φ is a continuous map from $\partial([0,1] \times B_2) \to f_{n_0+1,-M}$. Since $f_{n_0+1,-M}$ is homeomorphic to the unit ball in Λ_{n_0+1}, it is an absolute retract, which implies that Φ can be extended to a map

$$\tilde{\Phi}: [0,1] \times B_2 \to f_{n_0+1,-M}.$$

Set $\Gamma = \partial([0,1] \times B_2)$ and, $\forall n \geq n_{0+1}$ $Q_n = \{H_n \in C([0,1] \times B_2, \Lambda_n) \mid |H_n|_\Gamma = \tilde{\Phi}|_\Gamma = \Phi\}$.

We have that $Q_{n+1} \supset Q_n$ and, since $\tilde{\Phi} \in Q_{n_0+1}$, $Q_n \neq \emptyset$, $\forall n$.

We claim that $\forall n \geq n_0+1$, $\forall H \in Q_n$, we have that $H([0,1] \times B_2) \cap \partial B_1 \neq \emptyset$. Here B_1 denotes the ball of radius r in $(E_k)^\perp$ given by the local linking. Set $G = B_1 \times B_2$ and define, $\forall s \in [0,1]$, $F_s : G \to E$ as

$$F_s(x_1, x_2) = x_1 - H(s, x_2).$$

If the claim does not hold, $F_s(\partial B_1 \times B_2) \cap \{0\} = \emptyset$, $\forall s \in [0,1]$. We also have that $F_s(B_1 \times \partial B_2) \cap \{0\} = \emptyset$. In fact $F_{s_0}(x_1, x_2) = 0$ with $x_1 \in B_1$ and $x_2 \in \partial B_2$ implies $x_1 = H(s_0, x_2)$ and we know that for $s_0 = 0$ $H(0, x_2) = x_2 \in \partial B_2$, while $\forall s_0 \neq 0$ we have that $f(x_1) \geq 0$, $f(H(s_0, x_2)) < 0$, $\forall x_2 \in \partial B_2$.

So we have that $F_s(\partial G) \cap \{0\} = \emptyset$, $\forall s \in [0,1]$, so that $\deg(F_0, G, 0) = \deg(F_1, G, 0)$. But $\deg(F_1, G, 0) = \deg(x_1 - Re_{n_0+1}, G, 0) = 0$, while we have that $\deg(F_0, G, 0) = \deg(x_1 - x_2, G, 0) = (-1)^N$, contradiction which proves the claim.

From the claim, we immediately deduce that, setting

$$c_n = \inf_{H \in Q_n} \sup_{(t,x) \in [0,1] \times B_2} f_n(H(t, x_2))$$

one has $c_n \geq b > 0$ and also that $c_{n+1} \leq c_n \leq c_{n_0+1} < +\infty$.

Since the $(PS)^*$ condition is verified, we immediately deduce the existence of a critical value c of f. □

Proof of the Theorem. The theorem now follows by a simple application of Lemma 2.2.

In fact (f_1) follows from Lemma 1.1, $f_2)$ can be verified as in [L], (f_3) follows since Ω is bounded and finally (f_4) and (f_5) follow from Lemma 1.2. □

REFERENCES

[AC1] A. Ambrosetti and V. Coti Zelati, *Critical points with lack of compactness and singular dynamic systems.*, Annali di mathematica pura ed applicata (IV) **CIL** (1987), 237-259.

[AC2] A. Ambrosetti and V. Coti Zelati, *Solution with minimal period for Hamitonian system in a potential well*, Ann. inst. H. Poincare,Analyse nonlineaire **4:3** , (1987), 275-296.

[BR] A. Bahri and P.H. Rabinowitz, *A minimax method for a class of Hamiltonian systems with singular potentials*, J. Funct. Anal. **82** (1989), 412–428.

[B] V. Benci, *Normal modes of a Lagrangian system in a potential well*, Ann. inst. H. Poincare,Analyse nonlineaire **1** (1984), 294-1449.

[DG] M. Degiovanni and F. Giannoni, *Dynamical systems with Newtonian type potentials*, Ann.Scu. Nor. Sup. Pisa, **125** (1988), 467-494.

[GR] C. Greco, *Periodic solutions to second order Hamiltonian systems in an unbounded potential well*, Proc. Roy. Soc. Edinbourgh **105 A** (1987).

[G]   W.B. Gordon, *Conservative dynamical systems involving strong forces,*, Trans. Amer. Math. Soc. **204** (1975), 113-135.
[J]   Jiang Meiyue, *A remark on periodic solutions of singular Hamiltonian systems*. preprint.
[L]   Li Shujie, *Periodic solutions of nonautonomous second order systems with surperlinear terms*, preprint, S.I.S.S.A. (1990).
[R]   P.H. Rabinowitz, *Periodic solutions for some forced singular Hamiltonian systems,*, Festschrift fur Jurgen Moser (to appear).

NONERGODIC PROPERTIES OF SYSTEMS OF OSCILLATORS COUPLED THROUGH CONVEX EVEN POTENTIALS[†]

G.F. DELL'ANTONIO[*], B.D'ONOFRIO[**] and I. EKELAND[***]

[*] Dip. di Matematica, Univ. di Roma I
[**] SISSA, Trieste, Italy
[***] Univ. Paris IX, Dauphine
[†] Presented by B. D'Onofrio

1. This communication deals with finding elliptic periodic solutions to the boundary value problem

$$\dot{x} = JH'(x) \qquad x(o) = x(T) \qquad x \in R^{2n} \qquad 1.1$$

Here $H : R^{2n} \to R$ is convex, even and of class C^2 and J is the standard symplectic matrix.

I will illustrate the interest of the problem with some historical remarks. In the early fifties, in a famous computer simulation, Fermi, Pasta and Ulam studied numerically the behaviour of a chain of 64 harmonic oscillators with nonlinear coupling, descibed by the hamiltonian

$$H(q,p) \equiv \frac{1}{2}\sum_{1}^{64}[p_i^2 + q_i^2 - q_{i+1}^2 + \frac{\epsilon}{2}(q_i - q_{i+1})^4] \qquad 1.2$$

with suitable boundary conditions and ϵ small.

The objective was to verify Birkoff's ergodicity hypotheses. They observed on the contrary that, for initial data corresponding to the displacement of a single oscillator, the motion covered a small portion of the energy surface. Nowadays results of this type are interpreted in the context of KAM's theorem.

Our result shows that phenomena such as the one found by Fermi, Pasta and Ulam are present in a large class of strongly nonlinear oscillators with hamiltonian of the form

$$H(q,p) = \frac{1}{2}\sum_1^n p_i^2 + \sum_{i\neq j} V(q_i - q_j)$$

where the interaction potential energy V is convex and even.

2.

The system (1.1) is autonomous and therefore the motion takes place on surfaces of constant energy Σ_E. We show that if V is convex and even then on each Σ_E there exists at least one periodic solution $x_E(t)$ which is elliptic. This implies that generically there are subsets of Σ_E of positive Lebesgue measure which are left invariant by the hamiltonian flow. In fact, all eigenvalues of the Poincaré map π_E associated to $x_E(t)$ are on the unit circle, and under further (generic) conditions on V one can apply KAM's theorem and conclude about the existence on Σ_E of a set of positive Lebesgue measure fibered by invariant tori.

We shall prove:

THEOREM 2.1

Let Σ_E be a C^2 hypersurface in R^{2n} bounding a convex compact set symmetric under reflection about the origin. Then Σ_E carries at least one elliptic characteristic.

This implies in particular the existence of an elliptic periodic solution on each energy surface for the hamiltonian (1.4), as claimed.

We conclude this section providing some notation and basic assumptions. Consider the hamiltonian problem (1.1) and suppose that $x_T(t)$ is a periodic solution of period T. The tangent flow at x_T is described by the linear system

$$\dot{y} = A(t)y \qquad A(t) \equiv JH''(x_T(t)) \qquad \qquad 2.1$$

We shall denote by $R(t)$ the resolvent of (2.1).

DEFINITION 2.2

The system (2.1) is elliptic if all the eigenvalues of $R(T)$ have unit modulus.

Let Σ_E be the surface of energy E for the hamiltonian H and assume that $H'(x)$ never vanishes on Σ_E. Assume also that Σ_E is of class C^2 and bounds a convex compact set Ω with $0 \in \Omega$. Let $n(x)$ be the exterior normal to Σ_E at x; $n(x)$ is parallel to $H'(x)$ and therefore the closed characteristics of the system

$$\dot{x} = Jn(x) \qquad 2.2$$

coincide (geometrically) with the closed orbits of the hamiltonian flow. Remark that any other hamiltonian K, such that $K'(x)$ is parallel to $n(x)$ on Σ_E and never vanishes, gives rise to the same closed characteristics, and that the property of being elliptic is independent of the hamiltonian chosen.

For our analysis it is particularly convenient to choose for hamiltonian a function of the gauge j_Ω of Ω, defined by

$$j_\Omega(x) \equiv min\{\lambda : \frac{x}{\lambda} \in \Omega\} \qquad 2.3$$

As hamiltonian we choose

$$H_\alpha(x) \equiv (j_\Omega(x))^\alpha, \qquad 0 < \alpha < 2. \qquad 2.4$$

which is convex under our assumptions. We consider the boundary value problem

$$\dot{x} = JH_\alpha(x), \qquad H_\alpha = 1, \qquad \exists T : x(0) = x(T) \qquad 2.5$$

By a rescaling argument, (2.5) is equivalent to the fixed period problem

$$\dot{x} = JH_\alpha(x), \qquad x(0) = x(1) \qquad (H)$$

We are interested in finding elliptic solutions of problem (H); we note here that a first result in this direction was obtained by I.Ekeland [7], who proved:

THEOREM 2.3

Assume that Σ_E is (r,R)-pinched, with $R < \sqrt{2}\, r$. Then Σ carries at least one closed elliptic characteristic.

Recall that Σ_E is (r,R)-pinched if

$$R^{-2}|y|^2 \leq \frac{1}{2}(H''(x)y, y) \leq r^{-2}|y|^2 \qquad \forall x \in \Sigma_E \qquad 2.6$$

3. Proof of Theorem 3.1

We consider the boundary value problem

$$\dot{x} = JH'_\alpha(x), \quad x(1/2) = -x(0), \quad 0 < \alpha < 2 \qquad 3.1$$

On $X^\beta \equiv L^\beta([0, 1/2], R^{2n})$, $\alpha^{-1} + \beta^{-1} = 1$ we define the dual action functional

$$\psi(u) \equiv \int_0^{1/2} [\frac{1}{2}(u, \Pi u) + H^*_\alpha(-u)] dt \qquad 3.2$$

where H^*_α is the Fenchel-Legendre transform of H_α and $\Pi \equiv (Jd/dt)^{-1}$ is a bounded map from X^β to $H^1_{-1} \equiv \{u(t) : \dot{u}(t) \in L^2([0,1/2], R^{2n}), u(1/2) = -u(0)\}$.

One easily verifies that ψ is coercive and attains its minimum value at $u_0(t) \not\equiv 0$. Moreover $x_0(t) \equiv (Jd/dt)^{-1} u_0(t)$ is a nontrivial solution of (3.1).

By assumption $H(x) = H(-x)$ so that by (2.1) the flow tangent to $x_0(t)$ is *periodic of period* $1/2$.

We associate to (3.2) the quadratic form

$$Q_{-1}(w) \equiv \int_0^{1/2} [J\dot{w}, w) + ((H''_\alpha)(x_0(t))^{-1} J\dot{w}, J\dot{w})] dt \qquad 3.3$$

on H^1_{-1}. One has

PROPOSITION 3.1 [7]

If the functional (3.2) has a minimum at $u_0 \in L^\beta$ then Q_{-1} defined in (3.3) has index zero.

To state a Morse theory we complexify the problem. We introduce the spaces

$$H^1_\omega \equiv \{u : \dot{u} \in L^2([0, 1/2], C^{2n}), u(1/2) = \omega u(0), |\omega| = 1\} \qquad 3.5$$

and define Q^ω on H^1_ω by

$$Q^\omega(z) \equiv \int_0^{1/2} [(J\dot{z}, z)(t) + ((H''(x(t))^{-1} J\dot{z}, J\dot{z})] dt \qquad 3.6$$

Denote by $j(\omega) : S^1 \to N$ the index of Q^ω. We have already seen (Proposition 3.1) that under our assumptions one has $j(-1) = 0$.
We shall use

PROPOSITION 3.2

The discontinuity points of j_ω are the eigenvalues of $R(1/2)$ and $+1$

PROPOSITION 3.3

If ϵ is positive and sufficiently small, then $j(e^{i\epsilon}) \geq n$.

Proof

Set $z(t) = e^{2i\epsilon t} z_0$; then $z \in H_{e^{i\epsilon}}$. If ϵ is sufficiently small, one has

$$Q^{e^{i\epsilon}}(z) = 2\epsilon \int_0^{1/2} (iJz_0, z_0)dt + O(\epsilon^2) = \epsilon(iJz_0, z_0) + O(\epsilon^2)$$

Therefore for ϵ small the index of $Q^{e^{i\epsilon}}$ is not smaller than the index of the hermitian matrix iJ, i.e. not smaller than n.

We shall consider only the case in which -1 is not a Floquet multiplier of (3.3) (the general case is treated in [9]). Proposition 3.1 implies then $j(-e^{-i\epsilon}) = 0$ for ϵ small enough.

Consider the open arc $\{\omega : |\omega| = 1, Im\omega > 0\}$ and denote by n^+ the number of eigenvalues of $R(1/2)$ at which $j(\omega)$ increases moving counterclockwise (counting multiplicities), and by n^- the number of eigenvalues at which it decreases.

From Propositions 3.1, 3.2, 3.3 we conclude

$$n^- - n^+ \geq n \qquad 3.6$$

Since $n^- + n^+ \leq n$, one has $n^- = n$ so that all the Floquet multipliers of (3.3) have unit modulus.

Consider now $x_0(t)$ as solution of problem (H). Since $R(1) = (R(1/2))^2$ also the eigenvalues of $R(1)$ have unit modulus. This concludes the proof of Theorem 2.1.

REFERENCES

[1] Arnold, V.   *Méthodes mathématiques de la méchanique classique*
    Edition MIR, Moscou

[2] Clarke,F   *Periodic solutions of hamiltonian inclusions*
    J. Diff. Equations 40 (1981), 1-6

[3] Crocke,C., Weinstein,A   *Closed curves on convex hypersurfaces and periods of nonlinear oscillations*

Inv. Math. 64 (1981), 199-202

[4] Dell'Antonio,G.F., D'Onofrio,B.,Ekeland,I. *Stability from index estimates..*

to appear in Journal Diff. Eq.

[5] D'Onofrio,B., Ekeland,I. *Hamiltonian systems with elliptic periodic orbits*

Nonlinear Analysis T.M.A. 14 (1990) 11-21

[6] Ekeland,I *Une Théorie de Morse pour les systémes hamiltoniens convexes*

Ann. I.H.P. Analyse nonlinéaire 1 (1984) 19-78

[7] Ekeland,I. *An index theory for periodic solutions of convex hamiltonian systems*

Proceedings Symposia in Pure Math. 45 (1986) 395-423

[8] Fermi,E., Pasta,J., Ulam,S.

Collected papers of E.Fermi, Univ. of Chicago Press 1965, vol II

[9] Dell'Antonio,G.F., D'Onofrio,B.,Ekeland,I.

Les systémes hamiltoniens convexes et pairs ne sont pas ergodiques en general

Comptes Rendues de l'Academie des Sciences, Paris. (in press)

PERIODIC SOLUTIONS OF SECOND ORDER AUTONOMOUS SYSTEMS: EXISTENCE AND MULTIPLICITY FOR THE PRESCRIBED PERIOD PROBLEM

MARIO GIRARDI[*], MICHELE MATZEU[**]

(*) Dipartimento di Matematica Istituto G. Castelnuovo, Università di Roma "La Sapienza", P.le Aldo Moro n. 2, 00185 Roma
(**) Dipartimento di Matematica – Università di Roma "Tor Vergata", Via della Ricerca Scientifica, 00133 Roma

The aim of this paper is to present some recent results obtained by the authors about the problem of the existence of periodic solutions with a prescribed minimal period for second order autonomous systems of the type

$$\ddot{x} + V'(x) = 0 \qquad (V)$$

where $\ddot{x}$ is the second derivative with respect to the time of the unknown function $x : \mathbb{R} \to \mathbb{R}^N$ and V' is the gradient of a potential $V \in C^1(\mathbb{R}^N)$ which is supposed to have a *superquadratic* growth.

There are some results about this kind of problem which have been stated by many authors (also with a possible presence of a quadratic term added to V, which can appear in our results too), but, at our knowledge, all these results rely on the *convexity* assumption on V, except some partial results stated by Yiming Long.[6]

In[4] we have studied some cases where the *convexity* assumption is replaced by the *evenness* of V, together with a suitable technical condition on the second derivative V'', which will be mentioned after.

The solutions are found through the consideration of the critical points of the classical functional associated with (V), that is

$$I(v) = \frac{1}{2}\int_0^\tau |\dot{v}|^2 - \int_0^\tau V(v)$$

where τ can be either T, the prescribed period, or $T/2$.

Indeed we are able to give some different existence results, in the sense that the problem can be studied in some different functional framework spaces, for example starting from the space

$$E_1 = H_0^1(0, T/2; \mathbb{R}^N) \text{ (in this case one chooses } \tau = T/2)$$

or from the space

$$E_2 = H_{\text{odd}}^1(0, T; \mathbb{R}^N) = \{v \in H^1(0, T; \mathbb{R}^N) : v(0) = v(T)$$
$$v(t) = \sum_{\substack{k=2h+1 \\ h \in \mathbb{Z}}} a_k e^{ikt}, \ a_k \in \mathbb{C}, \ a_{-k} = \overline{a}_k\} \text{ (in this case one chooses } \tau = T)$$

and the critical points may be of Mountain Pass type on the whole space or on a suitable manifold, or constrained minimum points.

Actually, as a general fact, these different ways of constructing a critical point do not necessarily yield really different solutions to our problem: indeed, firstly, two different elements of the space may arise two identical solutions, in an obvious geometrical sense, i.e. they may belong to the same orbit (note, for example, that, under the everness assumption on V, for any possible solution x, *always* one indeed has a "pair of solutions", that is x and $-x$) secondly, it may be that a solution (what can happen for Mountain Pass solutions) has a minimal period *less* than T.

These are the reasons for which the true difficulty is to get a real multiplicity result.

Obviously, the alternative to the multiplicity result is the existence of only one solution having various different symmetry properties, each of one just derived by the specific variational method itself through which it has been obtained.

But now let us give the precise statements of the main results.

A first theorem can be stated by giving on V some "weak" assumptions of superquadratic growth, only localized at the origin and at infinity, in order to get a solution generated by a critical point of Mountain Press type.

THEOREM 1. (see [4]) Let $V \in C^2(\mathbb{R}^N)$ satisfy the following assumptions

$$V(x) \geq 0, \ \forall x \in \mathbb{R}^N, \ V(0) = 0 \qquad (V_1)$$
$$V(x) = o(|x|^2) \text{ as } x \to 0 \qquad (V_2)$$
$$\exists \beta > 2, \ R > 0 : V'(x) \cdot x \geq \beta V(x) > 0 \text{ if } |x| > R \qquad (V_3)$$
$$V(-x) = V(x) \quad \forall x \in \mathbb{R}^N \qquad (V_4)$$
$$V''(x)x \cdot x > V'(x) \cdot x \quad \forall x \in \mathbb{R}^N. \qquad (V_5)$$

Then, for any $T > 0$, there exists a T-periodic solution x of (V), with minimal period T and such that

$$x_1(t + T/2) = -x_1(T/2 - t) \quad \forall t \in [0, T/2] \tag{1}$$

$$x_1(t) \neq 0 \quad \forall t \neq kT/2 \quad k \in \mathbb{Z}. \tag{2}$$

Let us give now a sketch of the proof. One has to note that the functional

$$I(v) = \frac{1}{2} \int_0^{T/2} |\dot{v}|^2 - \int_0^{T/2} V(v)$$

considered on the space E_1, whose critical points coincide with the solutions of the problem

$$\begin{cases} \ddot{x} + V'(x) = 0 \\ x(0) = x(T/2) = 0 \end{cases} \tag{V_0}$$

verifies all the assumptions of the Mountain Pass theorem by Ambrosetti and Rabinowitz, namely:
- $I \in C^2$, $I(0) = 0$
- $I(v) > 0$ for $\|v\| = r$, sufficiently small
- $I(\bar{v}) < 0$ for some $\bar{v} \neq 0$
- the "Palais–Smale condition" is satisfied (i.e. $\{v_n\} \subset E_1$, $I'(v_n) \to 0$, $\{I(v_n)\}$ bounded imply the existence of a subsequence of $\{v_n\}$ strongly converging in E_1).

Then it is known that the number c defined as

$$c = \inf_{\gamma \in \Gamma} \max_{t \in [0,1]} I(\gamma(t))$$

where

$$\Gamma = \{\gamma \in C^0([0,1]; E_1) : \gamma(0) = 0, \quad \gamma(1) = \bar{v}\}$$

is a (positive) critical value of I, that is there exists $u_1 \in E_1$ such that $I(u_1) = c$, $I'(u_1) = 0$.

Moreover a well known result by Ekeland and Hofer[3] states that u_1 can be chosen in such a way that, if $i(u_1)$ is its Morse index, that is the dimension of the maximal subspaces of E_1 where the quadratic from Q associated with the second derivative of I at u_1, i.e.

$$Q(v) = \int_0^{T/2} |\dot{v}|^2 - \int_0^{T/2} V''(u_1) v \cdot v,$$

is negative definite, then one has

$$i(u_1) \leq 1$$

[Actually, one can show that, in the present case, one has $i(u_1) = 1$, due to the fact that
$$Q(u_1) < 0, \tag{3}$$
as a consequence of (V_5)].

At this point, following some similar arguments as those used by Benci and Fortunato[2] for a nonautonomous case, one can state the following property for u_1:
$$u_1(t) \neq 0 \quad \forall t \in (0, T/2). \tag{4}$$

Obviously (4) and (V_4) yield the possibility of exhibiting a T-periodic solution x_1, with minimal period T and satisfying (1), (2). Indeed one can consider the T-periodic extension on the whole real line of the function
$$\tilde{x}_1(t) = \begin{cases} u_1(t) \text{ for } t \in [0, T/2] \\ -u_1(T-t) \text{ for } t \in [T/2, T]. \end{cases}$$

The proof of (4) is based on the fact that, if $u_1(t^*) = 0$ for some $t^* \in (0, T/2)$, then the functions u_1^*, u_2^* given by
$$u_1^*(t) = \begin{cases} u_1(t) \text{ in } [0, t^*] \\ 0 \text{ in } [t^*, T/2] \end{cases} \qquad u_2^*(t) = \begin{cases} 0 \text{ in } [0, t^*] \\ u_1(t) \text{ in } [t^*, T/2] \end{cases}$$

would be two elements of E_1, Q-ortogonal each to the other and satisfying (as u_1),
$$Q(u_1^*) < 0, \quad Q(u_2^*) < 0.$$

A second type of solution, that is generated by a critical point of different type, can be obtained by strenghtening the superquadratic growth assumptions on V and property (V_5). Namely one has the following.

THEOREM 2. (see [4]) Let $V \in C^2(\mathbb{R}^N)$ satisfy (V_4) and let (V_1), (V_2), (V_3), (V_5) be replaced by the stronger assumptions

$$|V'(x)| \leq a_1 |x|^{\beta-1} \quad \forall x \in \mathbb{R}^N \tag{V_6}$$
$$V'(x) \cdot x \geq \beta V(x) > 0 \quad \forall x \in \mathbb{R}^N \setminus \{0\} \tag{V_7}$$
$$V''(x)x \cdot x \geq \mu V'(x) \quad \forall x \in \mathbb{R}^N, \quad \text{with} \quad \mu > 2. \tag{V_8}$$

Then, for any $T > 0$, there exists a T-periodic solution x_2 of (V), with minimal period T and such that
$$x_2(t + T/2) = -x_2(t) \quad \forall t \in [0, T/2]. \tag{5}$$

Let us give an idea of the construction of this second solution. It is related to the consideration of a suitable manifold containing all the critical points of I, which was used for the first time in the study of periodic solutions of Hamiltonian system by Ambrosetti and Mancini[1].

Let us choose as the functional space the following one:

$$E_2 = H^1_{\text{odd}}(0,T;\mathbb{R}^N), \quad \text{so } \tau = T \text{ in the definition of } I.$$

It is easy to verify that the symmetry assumption on V guarantees that the critical points of I on E_2 yield T–periodic solutions of (V), due to the fact that E_2 is a subspace of T–periodic H^1–functions which is invariant under V', thanks to (V_4). Moreover any critical point in E_2 verifies property (5).

Then it is possible to show that the critical points of I on E_2 coincide with the critical points of its restriction on the closed manifold

$$M_2 = \{v \in E_2 \smallsetminus \{0\} : \langle I'(v), v \rangle = 0, \text{ i.e. } \int_0^T |\dot v|^2 = \int_0^T V'(v) \cdot v\}$$

which is indeed regular and not empty, as easily follows from (V_8) (actually any $v \neq 0$ can be radially projected on a unique point of M_2). At this point one proves that I has indeed a minimum value on M_2 as a consequence of the facts that

- I is bounded from below on M_2, it is in fact positive;
- I satisfies the "Palais–Smale condition" on M_2,

and of a well known theorem by Rabinowitz[5].
Finally one can prove that the minimum point u_2 of $I_{/M_2}$ has the following further variational property

$$\int_0^T |\dot u|^2 = \min\Big\{\int_0^T |\dot v|^2 : v \in E_2 \smallsetminus \{0\}, \int_0^T V(sv) \geq \int_0^T V(su) \quad \forall s > 0,$$
$$\int_0^T V'(v) \cdot v \geq \int_0^T V'(u) \cdot u\Big\}. \tag{6}$$

Property (6) yields that the T–periodic extension of u_2 on the whole real line has in fact T as its minimal period, since if $T/_{2k+1}$ was a period of u_2 (note that the $T/2$–antiperiodicity of u_2 does not allow the even submultiples of T as possible periods of u_2), then the function

$$u^*(t) = u(t/_{2k+1})$$

would be an element of $E_2 \smallsetminus \{0\}$ such that

$$\int_0^T V(su^*) = \int_0^T V(su) \quad \forall s \geq 0$$

but
$$\int_0^T V'(u^*) \cdot u^* = \int_0^T V'(u) \cdot u$$

$$\int_0^T |\dot{u}^*|^2 = \frac{1}{2k+1} \int_0^T |\dot{u}|^2 < \int_0^T |\dot{u}|^2,$$

which is an absurdum.

We would note that this "constrained minimum solution" could also be obtained as a "Mountain Pass solution" on the whole space E_2, by the same argument of Theorem 1, for which the "weak" superquadratic growth conditions are sufficient. Nevertheless the argument to prove the minimality of the period T for the Mountain Pass solution in E_1 (that is in Theorem 1) cannot be used for this situation (in the space E_2, the "zero extensions" are not allowed!), and it is not clear how it can be adapted.

However these two different results can be considered the starting point for a multiplicity result. One can consider two different possibilities.

1. Consider, in the framework of assumptions of Theorem 2, the minimum point u_2 on the manifold M_2 and the minimum point u_1, on the analogous manifold M_1, in the space E_1: first, observe that the T–periodic solution x_1, associated with u_1, by the suitable extension mentioned in the proof of Theorem 1, has indeed minimal period T (for example, considering u_1 as a Mountain Pass on the whole space E_1); then, try to prove that x_1 and x_2 are geometrically distinct, showing that u_2 and the "extension" $\tilde{u}_1$ of u_1 in the space E_2 have different functional values or different Morse indexes.

2. Consider a fixed space, for example E_1, and consider a minimum point, say u_1, on the manifold M_1. Then one has two alternatives:

• u_1 is *not* an *isolated* minimum point: in this case, one has obviously *infinitely many different* solutions, corresponding to infinitely many different points "close" to u_1 in the E_1–norm;

• u_1 is an *isolated* minimum point. In this case, one can consider the other (isolated) minimum point $-u_1$, and the Mountain Pass critical point on the manifold M_1, say v_0, constructed by starting from the pair $(u_1, -u_1)$. It is obvious that v_0 is *different* from u_1, as $I(v_0) > I(u_1)$. Actually, the real problem is now to show that T is the *minimal* period of the solution x_0 associated with v_0 by the method described in the proof of Theorem 1. First of all, let us note that the Morse index of v_0 must be not greater than 2 (it is of "Mountain Pass type" on a manifold of codimension 1). This fact allows to state that T/k, with $k \geq 3$, cannot be a period of v_0, by using analogous arguments as those exhibited for the proof of the minimality of the period in Theorem 1.

The true difficulty is to exclude the period $T/2$.

At this moment, the authors have been able to yield this result, so the existence of 2 different T–periodic solutions of (V) with minimal period T, if the following situation (*) is verified:

(*) *The $T/2$–periodic solution $\overline{w}$ given by Theorem 1, that is constructed starting from the constrained minimum in $H_0^1(0, T/4)$, is not "$T/8$–symmetric", that is*

$$\overline{w}(t) \neq \overline{w}(T/4 - t), \quad t \in [0, T/8].$$

Let us give now only some very rough ideas how to exclude the period $T/2$ if (*) holds.

First of all, if x_0 had minimal period $T/2$, the construction itself of x_0 starting from v_0 would imply

$$v_0(T/4) = 0$$

and that the two functions

$$v_1(t) = \begin{cases} v_0(t), & t \in [0, T/4] \\ 0, & t \in [T/4, T/2] \end{cases} \qquad v_2(t) = \begin{cases} 0, & t \in [0, T/4] \\ v_0(t), & t \in [T/4, T/2] \end{cases}$$

would belong to $H_0^1(0, T/2)$ and satisfy the property

$$v_1(t) = -v_2(T/2 - t) \qquad \forall t \in [0, T/4].$$

Now one can check, as a consequence of a calculation based on the assumption on V'', that, if one considers the segment line S joining $v_0 - \varepsilon v_1$, and $v_0 + \varepsilon v_1$ (with $\varepsilon > 0$ sufficiently small), then the quadratic form associated to $I''(v_0)$ is *negative* on the projection $P_{M_1}(S)$ on M_1. In this situation, as a consequence of another known result by Ekeland and Hofer[3] about critical points of Mountain Pass type, the points $w_1 = P_{M_1}(v_0 - \varepsilon v_1)$ and $w_2 = P_{M_1}(v_0 + \varepsilon v_1)$ should belong to two different connected path components of the sublevel set

$$L = \{v \in M_1 : I(v) < I(v_0)\}.$$

The idea is just to try to construct a path γ on M_1 joining w_1 and w_2 in such a way that γ is contained in L, so obtaining a contradiction. Actually, condition (*) enables to exhibit this path γ, then to get the desired multiplicity result.

REFERENCES

[1] A. Ambrosetti, G. Mancini, Math. Ann., 255, pp. 405–421 (1981).
[2] V. Benci, D. Fortunato, Proceedings of the International Conference on Recent Advances in Hamiltonian systems, Univ. L'Aquila, 10–13/6/86, Ed. by G.F. Dell'Antonio, B.M. D'Onofrio.
[3] I. Ekeland, H. Hofer, Invent. Math., 81, pp. 155–188 (1985).
[4] M. Girardi, M. Matzeu, preprint.
[5] P. Rabinowitz, Minimax Methods in Critical Point Theory with Applications to Differential Equations, CBMS Reg. Conf. Ser. in Math., no. 65, Amer. Math. Soc., Providence R.I. (1986).
[6] Yiming Long, preprint.

EXISTENCE AND MULTIPLICITY RESULTS OF AN ELLIPTIC EQUATION IN UNBOUNDED DOMAINS.

MASSIMO GROSSI

Universitá degli Studi di Bari-Dip. di Matematica-Via Orabona 4
70125-Bari-Italy.

1. INTRODUCTION

In this paper we consider existence and multiplicity results for positive solutions of the problem

$$\begin{cases} -\Delta u + \lambda u = |u|^{p-1}u & \text{in } \Omega \\ u = 0 & \text{on } \partial\Omega \end{cases} \quad (0.1)$$

where $1 < p < \frac{N+2}{N-2}$, $N \geq 3$, $\lambda > 0$ and $\Omega \subset \mathbb{R}^N$ is an unbounded domain. Problems like (0.1) have been extensively studied in the last year; see for example Benci and Cerami[3], Bahri and Li[1], Berestycki and Lions[4], Cerami and Passaseo[5] and others.

It is well known that looking for solutions of (0.1) is equivalent to finding critical points of the functional $J : H_0^1(\Omega) \longrightarrow \mathbb{R}$

$$J(u) = \int_\Omega (|\nabla u|^2 + \lambda u^2) \quad (0.2)$$

constrained on the manifold $\Sigma^+ = \{u \in H_0^1(\Omega), u \geq 0 \text{ and } \int_\Omega |u|^{p+1} = 1\}$. The lack of compactness of the imbedding $i : H_0^1(\Omega) \hookrightarrow L^{p+1}(\Omega)$ when Ω is unbounded brings as a consequence that the Palais-Smale compactness condition is not satisfied for J on Σ^+ and this is an hard obstacle when one tries to apply the usual variational methods to find critical points. We recall that J satisfies the Palais-Smale condition at level c if for every $u_n \in \Sigma^+$ such that

$$\begin{cases} J(u_n) \mapsto c \\ J'(u_n) \mapsto 0 \quad \text{in } H^{-1}(\Omega) \end{cases} \tag{0.3}$$

then u_n admits a convergent subsequence.

The first existence results to (0.1) were obtained using symmetry assumptions on Ω to overcome these difficulties (see Bahri and Li[1] for the case $\Omega = \mathbb{R}^N$, Coffman and Marcus[6] for other kind of symmetry). Subsequently a careful analysis of the compactness failure (see Benci and Cerami[3] or Lions[10]) allowed to understand better the nature of the obstructions to the compactness. In fact in Lions[10] or Benci and Cerami[3], where it is assumed $\Omega = \mathbb{R}^N \setminus \overline{\omega}$ with ω bounded smooth open set, it has been showed that any Palais-Smale sequence either converges strongly to its weak limit or differs from it by one or more sequences which, after suitable translations, converge to a solution of

$$\begin{cases} -\Delta u + \lambda u = |u|^{p-1}u \quad \text{in } \mathbb{R}^N \\ u \in H^1(\mathbb{R}^N) \end{cases} \tag{0.4}$$

From this result it is possible to characterize the levels c of the functional J where the Palais-Smale condition fails. After that, using the topology of Ω, Benci and Cerami[3] obtained the existence of at least at solution to (0.1) in $\Omega = \mathbb{R}^N \setminus \overline{\omega}$ when meas(ω) is small enough. This result has been generalized by Bahri and Lions[2] at the case where ω is a general bounded domain of $\mathbb{R}^N$.

In section 1 we consider multiplicity results for (0.1) when $\Omega = \mathbb{R}^N \setminus \bigcup_{i=1}^{k} \overline{\omega}_i$, where ω_i are bounded smooth domain of $\mathbb{R}^N$ such that

$$\overline{\omega}_i \cap \overline{\omega}_j = \emptyset \quad \text{if} \quad i \neq j \quad \text{and} \quad \omega_i \subset \overline{B(x_i, \varepsilon_i)}, \ i = 1 \ldots k \qquad (0.5)$$

Here $B(x_i, \varepsilon_i)$ is the ball centered at the point x_i and radius ε_i. For such a domain we have the following

<u>Theorem 1</u> (See Grossi[8]). *Let us denote by $r_{ij} = |x_i - x_j|$. Then there exist real numbers $\delta_1, \delta_2(\varepsilon_1), \ldots, \delta_k(\varepsilon_1, \varepsilon_2, \ldots, \varepsilon_{k-1})$ and $\rho_1(r_{12}), \ldots, \rho_{k-1}(r_{ij})$, $j = 2, \ldots, k$, $i = 1, \ldots, k-1$ and $i < j$ such that for any $\varepsilon_1 < \delta_1, \ldots, \varepsilon_k < \delta_k$ and $|x_i - x_j| > \rho_{j-1}$ we have k positive solutions to (0.1) in $\Omega = \mathbb{R}^N \setminus \bigcup_{i=1}^{k} \overline{\omega}_i$.*

This theorem is proved by considering suitable min-max classes which provide critical values c_i, $i = 1, \ldots, k$ for the functional J satisfying

$$c_k < c_{k-1} < \ldots < c_2 < c_1 \qquad (0.6)$$

In section 2 we consider the case where $\partial\Omega$ is not necessarily bounded. In this case, the analysis of the behavior of the Palais-Smale sequences is more delicate. In fact it could happen that a Palais-Smale sequence differs from its weak limit by one or more sequences which, after suitable translations, converge to a function $u \in H_0^1(A)$ which solves

$$\begin{cases} -\Delta u + \lambda u = |u|^{p-1}u & \text{in } A \subset \mathbb{R}^N \\ u \in H_0^1(A) \neq H^1(\mathbb{R}^N) \end{cases} \qquad (0.7)$$

where A is a domain which depends on the geometry of Ω. Since in general we do not know the structure of the set of the solution to (0.7), we have imposed a geometrical condition (in terms of capacity) on Ω to avoid that this situation occurs. A particular case where problem (0.7) does not appear is provided

by $\Omega = \mathbb{R}^N \setminus \bar{\omega}$ with $\omega \subset G_\varepsilon = \{x = (x_1, x') \in \mathbb{R}^N, x_1 \in \mathbb{R}, x' \in \mathbb{R}^{N-1} :$
$|x'| < \dfrac{\varepsilon}{1 + |x_1|^\alpha}$, $\alpha > 0$ and $N \geq 4\}$. For such a domain we are able to construct a min-max procedure for the functional J which provides a solution to problem (0.1). At the end we remark that since we only require that $\omega \subset G_\varepsilon$, we can choose ω in such a way that Ω is a contractible domain.

1. THE MULTIPLICITY RESULT

We start this section by recalling some well known facts which will be used later. Let $\Omega = \mathbb{R}^N \setminus \bigcup_{i=1}^k \bar{\omega}_i$, where ω_i are bounded smooth domains satisfying (0.5),

$$\mu(\Omega) = \inf \left\{ \int_\Omega (|\nabla u|^2 + \lambda u^2) : u \in H_0^1(\Omega) \text{ and } \int_\Omega |u|^{p+1} = 1 \right\} \quad (1.1)$$

and

$$I = \inf \left\{ \int_{\mathbb{R}^N} (|\nabla u|^2 + \lambda u^2) : u \in H_0^1(\mathbb{R}^N) \text{ and } \int_{\mathbb{R}^N} |u|^{p+1} = 1 \right\}. \quad (1.2)$$

We have the following results:

<u>Proposition 2.</u> *The infimum I is achieved by a function $\bar{u}$ such that $I^{\frac{1}{p-1}}\bar{u}$ is the unique positive regular solution of (0.4) (modulo translations). Moreover $\bar{u}$ is spherically symmetric around some point in $\mathbb{R}^N$, $\frac{d}{dr}\bar{u}(r) < 0$ for $r>0$, r being the radial coordinate about that point and*

$$\lim_{r \mapsto \infty} r^{(N-1)/2} e^r \bar{u}(r) = \gamma > 0. \quad (1.3)$$

<u>Proof.</u> The claim of this proposition can be obtained just by combining the results of Bahri and Li[1], Gidas, Ni and Nirenberg[7], Rabinowitz[13] and Kwong[11].

<u>Proposition 3.</u> *We have that $\mu(\Omega) = I$ and $\mu(\Omega)$ is never achieved.*

<u>Proof.</u> See Benci and Cerami[3] or Grossi[8].

Proposition 4. *Let $u_n \in \Sigma^+(\Omega)$, $u_n \geq 0$, be a Palais-Smale sequence at level c for J. Then if*

$$I < c < 2^{\frac{p-1}{p+1}} I \tag{1.4}$$

u_n contains a strongly convergent subsequence.

Proof. See Benci and Cerami[3] (Lemma 3.1 and Corollary 3.4).

We prove Theorem 1 in the case $k = 2$ (see Grossi[8] for the general case $\Omega = \mathbb{R}^N \setminus \bigcup_{i=1}^{k} \overline{\omega}_i$). Hence

$$\Omega = \Omega(\varepsilon_1, \varepsilon_2, |x_1 - x_2|) = \mathbb{R}^N \setminus (\overline{\omega}_1 \cup \overline{\omega}_2)$$

with $\omega_1 \subset B(0, \varepsilon_1)$, $\omega_2 \subset B(x_2, \varepsilon_2)$, $\varepsilon_1, \varepsilon_2 < 1$ and such that $\overline{B(0, 2\varepsilon_1)} \cap \overline{B(x_2, 2\varepsilon_2)} = \emptyset$; for this domain we prove the existence of at least two solution for (0.1). As we have recalled in the Introduction, in Benci and Cerami[3] it has been proved that, if $\Omega = \mathbb{R}^N \setminus \overline{\omega}$ with meas(ω) small enough, there is a critical value c for the functional $\widetilde{J}(u) = \int_{\mathbb{R}^N \setminus \overline{\omega}} (|\nabla u|^2 + \lambda u^2)$ constrained on the manifold $\widetilde{\Sigma}^+ = \{u \in H_0^1(\mathbb{R}^N \setminus \overline{\omega}), u \geq 0 \text{ and } \int_{\mathbb{R}^N \setminus \overline{\omega}} |u|^{p+1} = 1\}$. With suitable modifications it is possible to prove an analogous result in the case $\Omega = \mathbb{R}^N \setminus (\overline{\omega}_1 \cup \overline{\omega}_2)$ and to obtain the existence of a critical value c_1 for J on Σ^+ such that

$$I < b < c_1 < 2^{\frac{p-1}{p+1}} I \tag{1.5}$$

where b is a constant which does not depend on ω_2. As in Benci and Cerami[3] we need to assume $\varepsilon_1 < \delta_1$ for some $\delta_1 > 0$. Our next aim is to obtain the existence of a second critical value for J by suitable hypothesis on ω_2. For this we fix $\omega_1 \subset B(0, \varepsilon_1)$ and set

$$\mathcal{F}_R = \{f \in C(\overline{B(x_2, R)}; \Sigma^+(\Omega)) \text{ such that } f_{|\partial B(x_2, R)} \equiv h\} \tag{1.6}$$

where

$$h(y)(x) = \frac{\psi(x)\bar{u}(x-y)}{(\int_\Omega |\psi(x)\bar{u}(x-y)|^{p+1})^{\frac{1}{p+1}}}, \quad y \in B(x_2, R), |x_2| > R^2 \quad (1.7)$$

Here $\bar{u}(x)$ is the function of $H^1(\mathbb{R}^N)$ spherically symmetric around the origin which achieves I (see Proposition 1) and $\psi(x)$ is the following function

$$\psi(x) = \begin{cases} \eta(\frac{x}{\varepsilon_1}) & \text{if } x \in B(0, 2\varepsilon_1) \\ \eta(\frac{x-x_2}{\varepsilon_2}) & \text{if } x \in B(x_2, 2\varepsilon_2) \\ 1 & \text{otherwise} \end{cases}$$

where $\eta : \mathbb{R}^N \mapsto [0,1]$ is a C^∞ radial function such that $\eta(x) = 0$ if $x \in B(0, \frac{3}{2})$ and $\eta(|x|) = 1$ if $|x| \geq 2$.

After that we define

$$c_2 = \inf_{f \in \mathcal{F}_R} \max_{x \in B(x_2, R)} J(f(x)) \quad (1.8)$$

We have the following lemmata (see Grossi[8])

<u>Lemma 5.</u> *If* $y_R \in \partial B(x_2, R), \varepsilon_2 < 1$ *and* $|x_2| > R^2$ *we have that* $\lim_{R \mapsto \infty} J(h(y_R))$
$= I$ *uniformly with respect to* ε_2.

<u>Lemma 6.</u> *For every* $y \in B(x_2, R)$, *with* $|x_2| > R^2$ *we have*

$$\lim_{\varepsilon_2 \mapsto 0} \lim_{R \mapsto \infty} J(h(y)) = I$$

uniformly with respect to y.

From Lemma 5 and the definition of c_2 it follows that there exists $\delta_2 > 0$ and $\tilde{R} > 0$ such that if $\varepsilon_2 < \delta_2$, $R > \tilde{R}$ we have

$$c_2 < b < c_1 < 2^{\frac{p-1}{p+1}} \quad (1.9)$$

Now, by a degree argument (see Benci and Cerami[3], Bahri and Li[1] or Grossi[8]), we have that

$$\max_{x \in B(x_2, R)} J(f(x)) \geq k > I \quad (1.10)$$

for some positive constant k and for every $f \in \mathcal{F}_R$. From (1.10) we deduce

$$c_2 \geq k > I \tag{1.11}$$

Then, from (1.9),(1.11) and Proposition 4 we have that the Palais-Smale condition holds at level c_2. Moreover, again from Lemma 5 and Lemma 6 it follows that, for every $R > \tilde{R}$

$$\max_{x \in B(x_2,R)} J(f(x)) > \max_{x \in \partial B(x_2,R)} J(f(x)) \quad \text{for every } f \in \mathcal{F}_R \tag{1.12}$$

Then, by the standard critical point theory (see Hofer[10]), we deduce that c_2 is a critical value for J on Σ^+ and $c_2 < c_1$ for $\varepsilon_1 < \delta_1$, $\varepsilon_2 < \delta_2(\varepsilon_1)$, R large; this proves Theorem 1 in the case $k = 2$.

2. AN EXISTENCE RESULT FOR SOME DOMAIN WITH UNBOUNDED BOUNDARY

In the previous section we have considered unbounded domains with bounded boundary. Now we want to establish an existence result for some domains of the type $\Omega = \mathbb{R}^N \setminus \overline{\omega}$ where ω can be un unbounded domain. We start giving a sufficient condition to hold also in this case the same result as in Proposition 3.

<u>Proposition 7.</u> *Let $\Omega = \mathbb{R}^N \setminus \overline{\omega}$, where ω is an open smooth set of $\mathbb{R}^N$. Let us suppose that the following condition holds:*

$$\begin{aligned}&\text{there exists } y_n \in \Omega \text{ such that } |y_n| \mapsto \infty \\ &\text{and } \operatorname{dist}(y_n, \partial \Omega) \mapsto \infty \text{ as } n \mapsto \infty.\end{aligned} \tag{2.1}$$

Then $\mu(\Omega) = I$ and $\mu(\Omega)$ is not achieved.

<u>Proof</u>(See Grossi[9]).

Our next aim is to characterize the critical values of the functional J where the Palais-Smale condition holds. To do this, we need to impose some geometrical conditions on Ω. We start with a definition.

Definition 8. Let $D \subset \mathbb{R}^N$ be a domain contained in a strip $E = \{x = (x_1, \ldots, x_N) \in \mathbb{R}^N : a < x_N < b, a, b \in \mathbb{R}\}$ and G a bounded smooth domain such that $\overline{G} \subset D$. Now we define the capacity of G with respect to D ($\mathrm{Cap}_D G$) as follows:

$$\mathrm{Cap}_D G = \inf\{\int_D |\nabla u|^2 : u \in H_0^1(D) \text{ and } u(x) \equiv 1 \text{ on } G \text{ almost everywhere}\} \tag{2.2}$$

Remark 9. In the Definition 8 we do not assume that D is bounded. On the other side, since D is contained in a strip, we can assume that $\|u\|_{H_0^1(D)} = \left(\int_D |\nabla u|^2\right)^{\frac{1}{2}}$ is a norm in $H_0^1(D)$ because of the Poincaré inequality. From this remark it is not difficult to prove that the infimum in (2.2) is achieved by a function $u \in H_0^1(D)$.

Now we can state the following compactness lemma

Lemma 10 (See Grossi[9]). Let $\Omega = \mathbb{R}^N \setminus \overline{\omega}$ be a smooth domain with $\omega \subset E = \{x = (x_1, \ldots, x_N) \in \mathbb{R}^N : a < x_N < b, a, b \in \mathbb{R}\}$. We assume that there exists $\tilde{\omega}$ with finite measure such that $\overline{\omega} \subset \tilde{\omega} \subset E$ and

$$\mathrm{Cap}_{\tilde{\omega}}(\omega \cap B(0, n)) \leq k \tag{2.3}$$

with k real constant not depending on $n \in \mathbb{N}$. Let u_n be a Palais-Smale sequence for the functional J in $\Sigma^+(\Omega)$. Then either u_n is relatively compact or there exist an integer $k \geq 1$, sequences $x_n^i \in \mathbb{R}^N$ and functions u^0 and $u^i, 1 \leq i \leq k$ such that

$$|x_n^i| \mapsto \infty, |x_n^i - x_n^j| \mapsto \infty \quad \text{as } n \mapsto \infty \quad \text{for } 1 \leq i \neq j \leq k \tag{2.4}$$

$$-\Delta u^0 + \lambda u^0 = c|u^0|^{p-1} u^0 \quad \text{in } \Omega, u^0 \in H_0^1(\Omega) \tag{2.5}$$

$$-\Delta u^i + \lambda u^i = c|u^i|^{p-1} u^i \quad \text{in } \mathbb{R}^N, u^i \in H^1(\mathbb{R}^N) \tag{2.6}$$

$$u_n - \left(u^0 + \sum_{i=1}^{k} u^i(x - x_n^i)\right) \mapsto 0 \quad \text{as } n \mapsto \infty \text{ strongly in } H^1(\mathbb{R}^N) \quad (2.7)$$

Remark 11. Recalling the analogous lemma of Benci and Cerami[3], we have that the levels of the functional J where the Palais-Smale condition does not hold are the same of the corresponding functional defined on $\Omega = \mathbb{R}^N \setminus \overline{C}$, where C is a bounded domain.

Remark 12. Let us consider the following set:

$$G_\varepsilon = \{x = (x_1, x') \in R^N : |x'| < \frac{\varepsilon}{1 + |x_1|^\alpha} \text{ for } \alpha > \frac{1}{N-3}, \varepsilon > 0 \text{ and } N \geq 4\}$$

We claim that the condition (2.3) holds with $\tilde{\omega} = G_{2\varepsilon}$. In fact it is possible to show (see again Grossi[9]) that (2.3) is equivalent to prove that there exists a function $\phi \in H_0^1(\tilde{\omega})$ such that $\phi \equiv 1$ almost everywhere on ω. Then if we consider

$$\phi(x) = \begin{cases} 1 & \text{if } x \in G_\varepsilon \\ 0 & \text{if } x \notin G_{2\varepsilon} \\ 2 - (1 + |x_1|^\alpha)\frac{|x'|}{\varepsilon} & \text{if } x \in G_{2\varepsilon} \setminus G_\varepsilon \end{cases}$$

we have, by easy calculations, that $\phi \in H_0^1(G_{2\varepsilon})$, $\phi \equiv 1$ on G_ε and then (2.3) holds.

Remark 13. What happens if (2.3) is not satisfied? In this case we have that some of the functions u^i of Lemma 9 are solutions of

$$\begin{cases} -\Delta u^i + \lambda u^i = |u^i|^{p-1} u^i & \text{in } A \\ u \in H_0^1(A) \end{cases} \quad (2.8)$$

with A possibly different from $\mathbb{R}^N$. For example if $\Omega = \mathbb{R}^3 \setminus \overline{\omega}_3$ with

$$\omega_3 = \{(x, y, z) \in \mathbb{R}^3 : x \in \mathbb{R}, 0 < y < 1, 0 < z < \frac{1}{1 + x^2}\}$$

then repeating the proof of Lemma 10 we get that either $A = \mathbb{R}^3$ or

$$A = \{(x,y,z) \in \mathbb{R}^3 : x \in \mathbb{R}, 0 < y - y_0 < 1, z = z_0\} \quad \text{for some } y_0, z_0 \in \mathbb{R}$$

In this case we do not know if there is any solution to (2.8). If there were we would obtain other critical values where the Palais-Smale condition is not satisfied.

After that, by considering a min-max argument very similar to that of the previous section, we obtain the following

<u>Theorem 14</u>(See Grossi[9]) . *Let us assume that $\Omega = \mathbb{R}^N \setminus \overline{\omega}$ with $\overline{\omega} \subset G_\varepsilon$ (G_ε as in the Remark 12). Then there exists a real number ε_0 such that if $\varepsilon < \varepsilon_0$ (and $N \geq 4$) there is at least a solution for (0.1) in Ω.*

For a more general result, involving a domain satisfying (2.3), we remind to Grossi[9].

REFERENCES

[1] A.Bahri-Y.Y.Li - *On a min-max procedure for the existence of a positive solution for certain scalar field equation in* $\mathbb{R}^N$. <u>Rev.Mat.Iberoamericana</u> (to appear).

[2] A.Bahri-P.L.Lions - *On the existence of a positive solution of semilinear elliptic equations in unbounded domains.* (to appear).

[3] V.Benci-G.Cerami - *Positive solutions of some nonlinear elliptic problems in exterior domains.* <u>Arch.RationalMech.Anal, 99</u>,p. **283-300** (1987).

[4] H.Berestycki-P.L.Lions - *Nonlinear scalar field equations I:existence of ground state.* <u>Arch.RationalMech.Anal., 82</u>,p. **313-346**(1983).

[5] G.Cerami-D.Passaseo - *Existence and multiplicity results of positive solution for nonlinear elliptic problems in exterior domains with "rich" topology.* <u>NonlinearAnalysisT.M.A., 18</u>,p. **109-119**(1992).

[6] C.V.Coffman-M.M.Marcus - *Existence theorems for superlinear elliptic Dirichlet problems in exterior domains.*

[7] B.Gidas-W.M.Ni-L.Nirenberg - *Symmetry of positive solutions of nonlinear elliptic equations in* $\mathbb{R}^N$. *MathematicalAnalysisandApplications, PartA, AdvancesinMathematicssupplementarystudies, Vol7A*,Academic Press (1981).

[8] M.Grossi -*Multiplicity results in semilinear elliptic equations with lack of compactness.* $Diff.andInt.Eq.$(to appear).

[9] M.Grossi -*An existence result for a semilinear equation in some domains with unbounded boundary.*(to appear)

[10] H.Hofer - *Variational and topological methods in partially ordered Hilbert space.* $Math.Ann.261$,**p. 493-514**(1982).

[11] M.K.Kwong - *Uniqueness of positive solutions of* $\Delta u - u + u^p = 0$. $\underline{Arch.\ RationalMech.Anal.}, 105$,**p. 243-266**(1989).

[12] P.L.Lions - *The concentration-compactness principle in the calculus of variation-The locally compact case-Part I.* $\underline{Ann.Ist.H.Poincaré.1}$, **p. 109-145**(1984).

[13] P.H.Rabinowitz -*Variational methods for nonlinear eigenvalue problems. Eigenvalue for nonlinear problems.* $\underline{(G.ProdiEd.),C.I.M.E.Ed.Cremo-nese,Roma}$, **p. 141-195**(1975).

MULTI-BUMP SOLUTIONS TO SCALAR CURVATURE EQUATIONS ON S^3, S^4 AND RELATED PROBLEMS

YANYAN LI[*]

[*] Department of Mathematics, Rutgers University,
New Brunswick, NJ 08903

Abstract

We show that for the prescribing scalar curvature problem on S^n ($n = 3, 4$), we can perturb any given positive function in any neighborhood of any given point on S^n such that for the perturbed function there exist as many solutions as we want to the prescribing scalar curvature equation on S^n ($n = 3, 4$). Critical exponent equations $-\Delta u = K(x)u^{\frac{n+2}{n-2}}$ in R^n ($n = 3, 4$) with $K(x) > 0$ being periodic in one of the variables are also studied and infinitely many positive solutions (Modulo translations by its periods) are obtained under some additional mild hypotheses on $K(x)$.

Let (S^n, g_0) be the standard n-sphere ($n \geq 2$). The prescribing scalar curvature problem is the problem of finding suitable conditions on $K(x)$, a positive function on S^n, such that $K(x)$ is the scalar curvature of a metric g on S^n conformally equivalent to g_0. Writing $g = u^{\frac{4}{n-2}} g_0$, this is equivalent to finding a positive function u on S^n which satisfies the following partial differential equation.

$$-\Delta_{g_0} u + \frac{n(n-2)}{4} u = \frac{n-2}{4(n-1)} K(x) u^{\frac{n+2}{n-2}}. \tag{1}$$

Where Δ_{g_0} denotes the Laplace-Beltrami operator associated with the metric g_0.

In the following, We will only quote a few results on the problem. For the details and more extensive references, please see [12], [3], [1], [11], [22], [4], [10], [23].... Theorem (Moser) : Let $K \in C^1(S^2)$ and $K(-x) = K(x) > 0$, then there is at least one solution to the prescribing scalar curvature problem.

Theorem (A. Chang & P.Yang for $n = 2$, A. Bahri & J.M. Coron for $n = 3$) : Let $K \in C^2(S^n)$ ($n = 2, 3$) being positive, with only non degenerate critical points $x_1, \ldots x_m$ of Morse index $k_1 \ldots k_m$. Assume further that

$$-\Delta K(x_i) \neq 0, \quad \forall \, i = 1, \ldots m.$$

If $\sum_{-\Delta K(x_i)>0} (-1)^{k_i} - (-1)^n \neq 0$, then (1) has at least one positive solution ($n = 3$). (If n=2, the equation is different. If we set $g = e^{2u}g_0$, then the equation is $-\Delta u + 1 = K(x)e^{2u}$.)

Theorem (A. Chang & P. Yang, we only state a weaker form) :

Let $K \in C^2(S^n)$ ($n \geq 4$) being positive, with only non degenerate critical points $x_1, \ldots x_m$ of Morse index $k_1 \ldots k_m$. Assume further that

$$-\Delta K(x_i) \neq 0 \quad \forall \, i = 1, \ldots m.$$

There exists $\epsilon_n > 0$, such that if $|K - 1|_{L^\infty} < \epsilon_n$ and

$$\sum_{-\Delta K(x_i)>0} (-1)^{k_i} - (-1)^n \neq 0,$$

then (1) has at least one positive solution.

Theorem (Z. Han for $n = 2$, D. Zhang & R. Schoen for $n = 3$) : In the "generic case", the "generalized Morse inequality" holds. In particular, in the "generic case", if $|\sum_{-\Delta K(x_i)>0} (-1)^{k_i} - (-1)^n|$ is large, then (1) has many solutions.

Our result concerning the prescribing scalar curvature problem is the following (see [14] and [15] for other results and more details.)

Theorem A: Suppose that $K(x) \in C^0(S^n)$ ($n = 3, 4$) and there exist some positive constants $K_1, K_2 > 0$ such that $K_1 \leq K(x) \leq K_2$ for any $x \in S^n$. Then for any $k = 1, 2, 3, \cdots$, $\epsilon > 0$, there exists $K_{\epsilon,k}(x) \in C^0(S^n)$, $\|K_{\epsilon,k} - K\|_{C^0(S^n)} < \epsilon$, such that, the equation

$$-\Delta_{g_0} u + \frac{n(n-2)}{4}u = \frac{n-2}{4(n-1)}K_{\epsilon,k}u^{\frac{n+2}{n-2}}$$

has at least k positive solutions.

Remark 1: The perturbation of $K(x)$ can be made in any small neighborhood of any given point on the sphere. We also know that the solutions we created are "m-bump solutions" ($m \geq 2$).

Remark 2: One can not expect to perturb any $K(x)$ near *any* point $\bar{x} \in S^n$ in C^1 sense to obtain the existence of solutions. This is evident if we take $K(x) = x^n + 2$ and $\bar{x}$ differente from the north pole and the south pole, since the perturbed function would still satisfy the Kazdan-Warner condition.

Let us look at a closely related problem in $R^n (n \geq 3)$.

$$\begin{cases} -\Delta u = K(x) u^{\frac{n+2}{n-2}} & \text{in } R^n \\ u > 0 \end{cases} \quad (2)$$

If $K(x)$ is "well behaved" at infinity, and if the solution u decays like $|x|^{2-n}$ at infinity. Then problem (2) is precisely the Kazdan-Warner problem.

For the equation (2), there have been many works. See [9], [13] ... and the references therein for details.

Before stating our results we introduce some notations.

Let E be the closure of $C_c^\infty(R^n)$ under the norm $\|u\|_E = (\int_{R^n} |\nabla u|^2)^{1/2}$. E is clearly a Hilbert space. We will simply use $\|\cdot\|$ to denote $\|\cdot\|_E$.

Theorem B: Suppose that $K(x) \in C^2(R^n)$ ($n = 3, 4$), satisfies (i) There exist some positive constants $K_1, K_2 > 0$, such that, $K_1 \leq K(x) \leq K_2$ for any $x \in R^n$. (ii) For some positive constant $T > 0$, $K(x_1 + lT, x_2, \cdots, x_n) = K(x_1, x_2, \cdots, x_n)$ for any integer l. (iii) $K_{max} = \max_{x \in R^n} K(x)$ is achieved and $K^{-1}(K_{max})$ has at least one bounded connected component. (iv) For $n = 4$, we further require that $\Delta K(x) = 0$ on those x in the bounded component with $\nabla K(x) = 0$. Then (2) has infinitely many positive solutions in E modulo translations by integer multiples of T in the first variable of $x \in R^n$.

Remark 3: Condition (iii) is quite natural due to the following nonexistence examples (basicaly the Kazdan-Warner obstruction).

Nonexistence Example: Suppose $K(x) \in C^1(R^3)$, $K(x)$ and $\nabla K(x)$ are bounded in R^3, $K_{x_2}(x)$ is nonnegative but not identically zero ($K_{x_2}(x)$ denotes the partial derivative of $K(x)$ in x_2-direction). Then the only solution of (1) in E is the trivial solution $u \equiv 0$.

Proof: Let u be any solution in E. Multiply (2) by u_{x_2} and integrate by parts. We obtain

$$\int_{R^3} K_{x_2}(x) u(x)^6 = 0.$$

The hypotheses on $K(x)$ implies that u is identically zero in an open set, hence $u \equiv 0$ by the well known unique continuation result.

Remark 4: Any smooth positive function $K(x)$ with at least one isolated maximum point (with zero Laplacian if $n = 4$) and periodic in at least one direction satisfies all the hypotheses above.

Remark 5: If the solution u of (2) belongs to E, then it follows from the standard elliptic theories that u decays like $|x|^{2-n}$ at infinity.

Theorem A and Theorem B follows from a more general result which we are not going to state here. See [14] and [15] for the details.

The proof consists of three main ingredients. One is some apriori estimates for the critical exponent equations (Proposition 1 and 2). The second is a slight modification of a minimax procedure as in Coti-Zelati and Rabinowitz ([7], [8]). See [20] for a closely related minimax procedure introduced earlier by Séré, as well as a related even earlier paper by Coti-Zelati, Ekeland and Séré ([6]). The third is the application of some compactness results along flow lines in dimension three due to Bahri and Coron ([1]) and our partial generalization to dimension four.

In the following we will only sketch the proof of Theorem B in the case $n = 3$. At the end of this note we will indicate the difference in the case $n = 4$.

We use the notation

$$E^+ = \{u \in E | u > 0 \text{ a.e. }\}.$$

We define also the Sobolev constant S_N by

$$S_N \equiv \min_{u \in E} \frac{(\int_{R^N} |\nabla u|^2)^{1/2}}{(\int_{R^N} u^{\frac{2N}{N-2}})^{\frac{N-2}{2N}}}, \quad N \geq 3.$$

$C_0(N)$ will always be used to denote some positive constant depending only on dimension N. The value of $C_0(N)$ may change here and there according to the context.

Proposition 1: Suppose that $K \in L^\infty(R^N \setminus B_1(0))$, $N \geq 3$, satisfies

$$0 < K_1 \leq K(x) \leq K_2 < \infty \quad \forall\, x \in R^N \setminus B_1(0) \tag{3}$$

for some positive constants K_1, K_2. Then there exists $\delta_1 = \delta_1(N, K_2) > 0$, $C_1(\delta_1, N, K_1, K_2) > 0$, such that, for any positive solutions of

$$-\Delta u = K(x) u^{\frac{N+2}{N-2}}, \quad |x| \geq 1 \tag{4}$$

with $\nabla u \in L^2(R^N \setminus B_1(0))$, $u \in L^{\frac{2N}{N-2}}(R^N \setminus B_1(0))$ and

$$\int_{|x| \geq 1} |\nabla u|^2 + \int_{|x| \geq 1} u^{\frac{2N}{N-2}} \leq \delta_1.$$

We have

$$u(x) \leq C_1(\delta_1, N, K_1, K_2) |x|^{2-N}, \quad \forall\, |x| \geq 2.$$

Proposition 2: Let $0 < K_1 \leq K_2 < +\infty$ be two constants, $\delta_1 > 0$ be the constant in Proposition 1. Then for any $2 < l_1 < l_2 < +\infty$, there exists

a positive constant, $R_1 = R_1(K_1, K_2, \delta_1, l_1, l_2) > l_2$, such that, for any positive solution of
$$-\Delta u = K(x)u^{\frac{N+2}{N-2}}, \quad 1 < |x| < R_1$$
with
$$K_1 \leq K(x) \leq K_2, \quad 1 < |x| < R_1$$
and
$$\int_{1<|x|<R_1} |\nabla u|^2 + \int_{1<|x|<R_1} u^{\frac{2N}{N-2}} \leq \delta_1,$$
we have
$$u(x) \leq \frac{2C_1}{|x|^{N-2}}, \quad \forall \; l_1 \leq |x| \leq l_2,$$
where $C_1 = C_1(\delta_1, N, K_1, K_2)$ is the constant in Proposition 1. In particular, it is independent of l_1 and l_2.

Let
$$I(u) = \frac{1}{2} \int_{R^N} |\nabla u|^2 - \frac{N-2}{2N} \int_{R^N} K(x)|u|^{\frac{2N}{N-2}}, \quad u \in E, \tag{5}$$
and
$$I^b = \{u \in E | I(u) \leq b\},$$
$$I_a = \{u \in E | I(u) \geq a\},$$
$$\mathbf{K} = \{u \in E | I'(u) = 0\},$$
$$\mathbf{K^+} = \{u \in E^+ | I'(u) = 0\}.$$

It is not difficult to check that $I \in C^2(E, R)$.

For $\epsilon > 0$, we define $V(2, \epsilon)$ as the following.

$u \in V(2, \epsilon)$ if $u \in E$ and there exists $\alpha_1, \alpha_2 \in R$, $x = (x_1, x_2), x_1, x_2 \in R^N$, $\lambda = (\lambda_1, \lambda_2), \lambda_1, \lambda_2 > 0$, such that,
$$\lambda_1, \lambda_2 > \frac{1}{\epsilon},$$
$$\frac{\lambda_1}{\lambda_2} + \frac{\lambda_2}{\lambda_1} + \lambda_1\lambda_2|x_1 - x_2|^2 > \frac{1}{\epsilon},$$
$$|\alpha_i - \frac{1}{K(x_i)^{\frac{N-2}{4}}}| < \epsilon, \quad i = 1, 2.,$$
$$\|u - \varphi(\alpha, x, \lambda)\| < \epsilon,$$
where
$$\varphi(\alpha, x, \lambda) := \alpha_1\delta(x_1, \lambda_1) + \alpha_2\delta(x_2, \lambda_2),$$
$$\delta(x, \lambda)(y) := (N(N-2))^{\frac{N-2}{4}} \left(\frac{\lambda}{1+\lambda^2|y-x|^2}\right)^{\frac{N-2}{2}}.$$

It is well known that
$$-\Delta\delta(x,\lambda) = \delta(x,\lambda)^{\frac{N+2}{N-2}} \quad \text{in } R^N,$$
for any $x \in R^N, \lambda \in (0, \infty)$.

Proposition 3: There exists $\epsilon_0 \in (0,1)$, such that, for any $0 < \epsilon \leq \epsilon_0$, $u \in V(2,\epsilon)$,
$$\min_{(\alpha,x,\lambda) \in B_{4\epsilon}} \|u - \varphi(\alpha, x, \lambda)\| \tag{6}$$
has a unique minimum, and it is achieved in $B_{2\epsilon}$. Where
$$B_\epsilon = \{(\alpha, x, \lambda) | x = (x_1, x_2), x_1, x_2 \in R^N, \frac{1}{2K_2^{\frac{1}{4}}} \leq \alpha_i \leq \frac{2}{K_1^{\frac{1}{4}}}$$
$$\lambda = (\lambda_1, \lambda_2), \lambda_1, \lambda_2 \geq \frac{1}{\epsilon}, \frac{\lambda_1}{\lambda_2} + \frac{\lambda_2}{\lambda_1} + \lambda_1 \lambda_2 |x_1 - x_2|^2 \geq \frac{1}{\epsilon}\}.$$

Let
$$\Gamma = \{g \in C([0,1], E) | g(0) = 0, g(1) \in I^0 \setminus \{0\}\},$$
$$c = \inf_{g \in \Gamma} \max_{\theta \in [0,1]} I(g(\theta)).$$

It is obvious that $c > 0$.

Theorem B will be proved by the contradiction argument. Therefore *from now on we always assume* that for some $\alpha > 0$,
$$\sup\{\|u\|_{L^\infty} | u \in \mathbf{K}^+ \cap (I_c^{c+\alpha} \cup I_{2c-\alpha}^{2c+\alpha})\} \leq C(K, \alpha). \tag{7}$$

Lemma 1: $c = \frac{1}{N}(K_{max})^{\frac{2-N}{2}}(S_N)^N$.
For $A > 0$, set
$$S_A = \{y = (y_1, y_2, y_3) \in R^3 | |y_2| + |y_3| \leq A\}$$

Proposition 4: Suppose that $K(x)$ satisfies (i)-(iii) with $N = 3$. Then there exists some constant $\epsilon_1 \in (0, \epsilon_0)$, such that, for any $0 < \epsilon \leq \epsilon_1$, $A > 1$, there exists $\delta_1(\epsilon, A) > 0$, such that
$$\inf\{\|I'(u)\| | u \in V(2,\epsilon) \setminus V(2, \frac{\epsilon}{2}), x_1(u), x_2(u) \in S_A,$$
$$|x_1(u) - x_2(u)| \geq 10, |I(u) - 2c| \leq \epsilon_1\} \geq \delta_1(\epsilon, A).$$

Remark 6: Proposition 4 holds also in R^N for $N \geq 4$.

Under the hypotheses (i)-(iv), we can find a bounded open set O in R^3, which is a union of finite open balls, satisfying:
$$O \cap K^{-1}(K_{max}) \neq \phi \tag{8}$$

$$\max_{x \in \partial O} K(x) < K_{max} \tag{9}$$

Without loss of generality (we can always scale x variable), we assume that the diameter of O is less than one tenth, namely,

$$\sup\{|x-y| | x, y \in O\} < \frac{1}{10}. \tag{10}$$

It follows from (9) that there exists $\epsilon_2 \in (0, \min\{\epsilon_0, \epsilon_1\})$, such that,

$$\inf\{I(u) | u \in V(2, \epsilon_2), \{x_1(u), x_2(u)\} \cap \{\cup_{l=-\infty}^{+\infty}\{\partial O + (lT, 0, 0)\}\} \neq \phi\} \geq 2c + \frac{1}{C}. \tag{11}$$

Where $C = C(K) > 0$ depends on $K_{max} - \max_{x \in \partial O} K(x) > 0$. Choose $\epsilon_3 \in (0, \min\{\epsilon_2, \frac{1}{C}\})$.

Let
$$\Gamma_2 = \{G = g_1 + g_2 | g_1, g_2 \text{ satisfies } (72)-(75)\}$$

$$g_i \in C([0,1]^2, E), \quad i = 1, 2, \tag{12}$$

$$g_1(0, \theta_2) = g_2(\theta_1, 0) = 0, \quad 0 \leq \theta_1, \theta_2 \leq 1, \tag{13}$$

$$g_1(1, \theta_2), g_2(\theta_1, 1) \in I^0 \setminus \{0\}, \quad 0 \leq \theta_1, \theta_2 \leq 1, \tag{14}$$

$$(\cup_{\theta=(\theta_1,\theta_2)\in[0,1]^2} supp\, g_1(\theta)) \cap (\cup_{\theta=(\theta_1,\theta_2)\in[0,1]^2} supp\, g_2(\theta)) = \phi. \tag{15}$$

Remark 7:

$$\{g_1(\theta_1) + g_2(\theta_2) | g_1, g_2 \in \Gamma, (75) \text{ holds }\} \subset \Gamma_2.$$

Set
$$b_2 = \inf_{G \in \Gamma_2} \max_{\theta \in [0,1]^2} I(G(\theta)).$$

Proposition 5: $b_2 = 2c$.

In the following, we show that under the contradiction hypotheses (7) we are able to construct some element $H \in \Gamma_2$ with

$$\max_{\theta \in [0,1]^2} I(H(\theta)) < 2c$$

which is a contradiction to $b_2 = 2c$.

We achieve this by (roughly speaking) the following three steps.

Step 1: We construct $G \in \Gamma_2$ with

$$\max_{\theta \in [0,1]^2} I(G(\theta)) \leq 2c + \epsilon.$$

G is constructed explicitly.

Step 2: We follow the negative gradient flow of I to deform G to U with
$$\max_{\theta \in [0,1]^2} I(U(\theta)) \leq 2c - \epsilon$$

However, U does not have to be in Γ_2 any more. Here we have used the deep result of Bahri and Coron on the compactness along flow lines in dimension three ([1]). In four dimention, we have some partial generalization which is enough to carry out this step.

Step 3: Using Proposition 2, we are able to modify U to obtain $H \in \Gamma_2$ with
$$\max_{\theta \in [0,1]^2} I(H(\theta)) \leq 2c - \frac{\epsilon}{2}.$$

Acknowledgment: We thank Professor Nirenberg, Professor Brezis and Professor Bahri for their constant encouragement and many stimulating conversations.

References

[1] A. Bahri and J.M. Coron, The scalar-curvature problem on the standard three-dimensional sphere. J. of Func. Anal., Vol. 95, No. 1(1991), pp. 106-172.

[2] H. Brezis and L. Nirenberg, Positive solutions of nonlinear elliptic equations involving critical Sobolev exponents. Comm. Pure. Appl. Math., 36(1983), pp. 437-477.

[3] S.Y. Chang and P. Yang, Conformal deformations of metrics on S^2, J. Diff. Geom., 27(1988), pp. 256-296.

[4] S.Y. Chang and P. Yang, A perturbation result in prescribing scalar curvature on S^n, preprint.

[5] W.X. Chen and W. Ding, Scalar curvature on S^2. Trans. Amer. Math. Soc., 303(1987), pp. 365-382.

[6] V. Coti Zelati, I. Ekeland and E. Séré, A variational approach to homoclinic orbits in Hamiltonian systems, Math. Ann., 288(1990), pp. 133-160.

[7] V. Coti Zelati and P.H. Rabinowitz, Homoclinic orbits for second order Hamiltonian systems possessing superquadratic potentials, preprint.

[8] V. Coti Zelati and P.H. Rabinowitz. Homoclinic type solutions for a semi-linear elliptic pde on R^n, preprint.

[9] W. Ding and W.-M. Ni, On the elliptic equation $\Delta u + Ku^{\frac{n+2}{n-2}} = 0$ and related topics. Duke Math. J., 52(1985), pp. 485-506.

[10] J. Escobar, Conformal deformation of a Riemannian metric to a scalar flat metric with constant mean curvature on the boundary, preprint.

[11] Z.C. Han, Prescribing Gaussian curvature on S^2, Duke Math. J., Vol. 61, No.3(1990), pp.679-703.

[12] J. Kazdan and F. Warner, Existence and conformal deformation of metrics with prescribed Gaussian and scalar curvature. Ann. of Math. 101(1975), pp. 317-331.

[13] Y. Li and W.-M. Ni, On conformal scalar curvature equations in R^n. Duke Math. J., 57(1988), pp. 895-924.

[14] Y.Y. Li, On $-\Delta u = K(x)u^5$ in R^3, to appear in Comm. Pure Appl. Math..

[15] Y.Y. Li, Prescribing scalar curvature on S^3, S^4 and related problems, preprint.

[16] Y.Y. Li, On prescribing scalar curvature problem on S^3 and S^4. C. R. Acad. Sci. Paris, t. 314(1992), Série I, pp. 55-59.

[17] P.L.Lions, The concentration-compactness principle in the Calculus of Variations. The locally compact case, part 1 and part 2. Ann. Inst. Henri Poincare, Analyse non linear, Vol. 1, No. 2(1984), pp. 109-145 and pp. 223-283.

[18] J. Moser, On a nonlinear problem in differential geometry, Dynamical systems (M. Peixoto, ed.) Academic Press, New York, 1973.

[19] Wei-Ming Ni, On the elliptic equation $\Delta u + K(x)u^{\frac{n+2}{n-2}} = 0$, its generalization and application in geometry, Indiana Univ. Math. J. 31(1982), pp. 493-529.

[20] E. Séré, Existence of infinitely many homoclinic orbits in Hamiltonian systems, Math. Z., 209 (1992), pp. 27-42.

[21] R. Schoen, Lecture notes of special topics in geometry courses in Stanford University and New York University, 1988 and 1989.

[22] R. Schoen and D. Zhang, in preparation.

[23] D. Zhang, , New results on geometric variational problems, thesis, Stanford University, 1990.

GEOMETRY AND TOPOLOGY OF THE BOUNDARY IN SOME SEMILINEAR NEUMANN PROBLEMS

GIOVANNI MANCINI

Dipartimento di Matematica, Università di Bologna
Piazza di Porta San Donato 1-40126 Bologna, Italia

1. INTRODUCTION

Let $\Omega \subset \Re^N, N \geq 3$, be a smooth bounded domain, with boundary $\partial\Omega = \Gamma_0 \cup \Gamma_1$, where Γ_0 and Γ_1 are disjoint smooth submanifolds of dimension $N-1$; ν will denote the normal at boundary points. Let $p \in (2, 2^*], 2^* = \frac{2N}{N-2}, \lambda > 0$ be given.
We will consider the following problem

$$\begin{aligned} -\Delta u + \lambda u &= u^{p-1} \quad \text{in } \Omega \\ u &> 0 \quad \text{in } \Omega \\ u &= 0 \quad \text{in } \Gamma_0 \\ \frac{\partial u}{\partial \nu} &= 0 \quad \text{in } \Gamma_1 \end{aligned} \quad (1.1)$$

In contrast with the analogous Dirichlet problem (e.g.: uniqueness, in case $p < 2^*$, non existence in case $p = 2^*$, if Ω is a ball) we have the following results:

Theorem 1.1 ([10], see also [13]) Let $p < 2^*$. Then, for λ large, (1.1) has at least $1 + cat(\partial\Omega)$ non constant solutions.

Theorem 1.2 ([1], see also [12]). Let $p = 2^*$, $\Gamma_1 = \partial\Omega$. Then, for λ large, (1.1) possesses a non constant solution.

Actually, (1.1) is solvable under the weaker assumption (in case $p = 2^*$)

$$\Gamma_1^+ := \{x \in \Gamma_1 : H(x) > 0\} \quad \text{is not empty} \tag{1.2}$$

Here $H(x)$, $x \in \partial\Omega$, denotes mean curvature.
As noticed in [1], one cannot drop, in general, assumption (1.2). Nevertheless, one might have existence without assumption (1.2): this is the case if Ω is an annulus and Γ_1 is the inner boundary. As far as we know, no general result, in case Γ_1^+ is empty, is known. However, the geometric assumption on Γ_1 might be replaced by a topological one. A partial result in this direction is the following:

Theorem 1.3 *(see [2]) Let $p = 2^*$. Assume Γ_1 has at least two, suitably close, connected components. Then (1.1) has at least one nonconstant solution, for λ large.*

A result similar to Theorem 1.1 can be obtained in the critical case as well, with curvature assumptions coming into picture. Let us denote

$$\operatorname{cat}\Gamma_1^+ = \sup\{\operatorname{cat}_{\Gamma_1^+} K : K \subset \Gamma_1^+, K \text{ compact}\}.$$

Theorem 1.4 *([2], see also [14]). Let $p = 2^*$. Then (1.1) has at least $\operatorname{cat} \Gamma_1^+$ non constant solutions, provided λ is large enough.*

Remark 1.1 *Ground state solutions of (1.1) are known to concentrate around boundary points as λ goes to infinity (see [11] for $p < 2^*$, and [3], in case $p = 2^*$). In [3] it is shown that concentration points maximize the mean curvature H. This is the starting point, in [3], to obtain a (different kind of) multiplicity result for (1.1).*

2. BASIC ESTIMATES, A PROOF OF THEOREM 1.4 AND FURTHER REMARKS

In this section $p = 2^*$. We will use the following notations:
$|u|_p = (\int_\Omega |u|^p)^{\frac{1}{p}}$ for $u \in H^1(\Omega)$
$\| u \|^2 = \int_\Omega |\nabla u|^2$ for $u \in H^1(\Omega)$.
S will denote the best Sobolev constant given by
$S := \inf\{\int_\Omega |\nabla u|^2 : |u|_p = 1, u \in H_0^1(\Omega)\}$
We will write $\hat{S} := \frac{S}{2^{\frac{2}{N}}} = \inf\{\int_{\Re_+^N} |\nabla u|^2 : \int_{\Re_+^N} |u|^p = 1, u \in H^1(\Re_+^N)\}$
where $\Re_+^N$ is a halfspace in $\Re^N$.

S is attained at $U = \frac{c_N}{(1+|x|^2)^{\frac{N-2}{2}}}$. Let V be the extremal function for $\hat{S}$, $V = \alpha U$, with $|V|_p = 1$.
For $\epsilon > 0$, $\sigma \in \Re^N$, set

$$V_{\epsilon,\sigma}(x) = \epsilon^{-\frac{N-2}{2}} V(\frac{x-\sigma}{\epsilon})$$

$V_{\epsilon,\sigma}$ are still extremal functions.
Let $E_\lambda(u) = \int_\Omega |\nabla u|^2 + \lambda |u|^2$. As in [7], one can show that the inequality

$$S_\lambda := \inf\{E_\lambda(u) : |u|_p = 1, \quad u \in H^1(\Omega), u = 0 \quad \text{on } \Gamma_0\} < \hat{S} \quad (2.1)$$

implies S_λ is achieved.
In [1] the following basic estimate is established

$$E_\lambda(V_{\epsilon,\sigma}) = \hat{S} - A_N H(\sigma)\epsilon + a_N \lambda o(\epsilon) + h.o.t. \quad (2.2)$$

for some A_N, a_N positive constants.
Clearly, (2.2) implies (2.1), and hence (1.1) is solvable.
Theorem 1.1 and 1.4 fit in the framework of a variational principle due to Benci-Cerami ([4]). The following version is adapted from [5].

Theorem 2.1 *Let $K \subset \Gamma_1^+$ be a given compact set. Assume there are continuous maps*

$$\Phi : K \to \{u > 0 : \quad |u|_p = 1, \quad E_\lambda < \hat{S}\}$$

$$\beta : \{u > 0 : \quad |u|_p = 1, \quad E_\lambda \leq \hat{S}\}$$

such that $\beta \circ \Phi$ is homotopic to the inclusion map $j : K \to \overline{\Gamma_1^+}$. Then E_λ has at least $\mathrm{cat}_{\overline{\Gamma_1^+}} K$ dinstinct critical points in $\{u > 0 : |u|_p = 1, \quad E_\lambda < \hat{S}\}$.

In order to apply Theorem above, we first observe that, for every $\lambda > 0$, we have, in view of (2.2),

$$\sup\{E_\lambda(V_{\epsilon,\sigma}) : \sigma \in K\} < \hat{S} \quad \text{for every} \quad K \subset \Gamma_1^+, \quad K \text{ compact}$$

provided $\epsilon = \epsilon(\lambda)$ is small enough.
To continue, let us consider

$$\beta(u) := \int_\Omega x|u|^p$$

A simple argument (see Lemma 4.2 in [2]), based on P.L.Lions' concentration lemma (see Lemma I.1 in [9]), shows that

$$E_\lambda(u_\lambda) \leq \hat{S}, \lambda \to +\infty \Rightarrow \exists \lambda_j \to +\infty, \exists \sigma \in \partial\Omega \quad \text{such that}$$
$$|\nabla u_{\lambda_j}|^2 \rightharpoonup \hat{S}\delta_\sigma, \quad |u_\lambda|^p \rightharpoonup \delta_\sigma \quad \text{(weak convergence of measures)} \quad (2.3)$$

and hence

$$\forall \delta > 0 \exists \lambda_\delta : \lambda \geq \lambda_\delta, E_\lambda(u) \leq \hat{S} \Rightarrow \beta(u) \in N_\delta(\partial\Omega)$$

where $N_\delta(\partial\Omega)$ is a δ-neighborhood of $\partial\Omega$.
More refined estimates, taken from [3], yield:

$$\forall \delta > 0 \exists \lambda_\delta : \lambda \geq \lambda_\delta, E_\lambda(u) \leq \hat{S} \Rightarrow \beta(u) \in N_\delta(\Gamma_1^+) \quad (2.4)$$

This gives Theorem 1.4, in view of Theorem 2.1. To see (2.4) we just need the following:

Lemma 2.1 *Let $E_{\lambda_n}(u_n) < \hat{S}, \lambda_n \to \infty$. Then $u_n \rightharpoonup 0$ and there is $\sigma_n \in \partial\Omega, \epsilon_n \to 0$ such that*

$$\int_\Omega |\nabla(u_n - V_{\epsilon_n,\sigma_n})|^2 \to 0$$

Lemma 2.2 *([3], Lemma 2.3). Let $\epsilon_n \to 0, \sigma_n \in \partial\Omega, \sigma_n \to \sigma, u_n \rightharpoonup 0$ be such that*

$$\int_\Omega |\nabla(u_n - V_{\epsilon_n,\sigma_n})|^2 \to 0$$

Then
$$E_{\lambda_n}(u_n) \geq \hat{S} - A_N H(\sigma_n)\epsilon_n + a_N \lambda_n o(\epsilon_n) + h.o.t.$$

Now, if u_n is as in Lemma 2.1, $H(\sigma_n) > 0$ by Lemma 2.2, and $|\nabla u_n|^2 \rightharpoonup \hat{S}\delta_\sigma$ weakly in the sense of measures. This yields (2.4).
The proof of Lemma 2.1 is based on a blowup technique and the concentration lemmas of P.L.Lions.
One first chooses ϵ_n, σ_n such that

$$\int_{B_{\epsilon_n}(\sigma_n) \cap \Omega} |u_n|^p = \sup\{\int_{B_{\epsilon_n}(x) \cap \Omega} |u_n|^p : x \in \Omega\} = \delta, \quad \delta \in (0,1) \quad \text{given.}$$

By (2.3) it follows that $\epsilon_n \to 0, \sigma_n \to \sigma \in \partial\Omega$.

Now, let $v_n(x) = \epsilon_n^{\frac{N-2}{2}} u_n(\epsilon_n x + \sigma_n)$, $x \in \Omega_n := \frac{\Omega - \sigma_n}{\epsilon_n}$. Following the arguments in [8] (Theorem I.1) we see that $\rho_n := \chi_n |v_n|^p$ is tight: clearly ρ_n does not vanish and dychotomy can be ruled out in a standard way. Thus, $\forall \epsilon > 0 \exists R : \int_{B_R} |v_n|^p \geq 1 - \epsilon$. Also, one can see, with similar estimates, that $|\nabla v_n|^2$ is tight as well and this in turn implies that Ω_n tends to a half space. Finally, after suitably extending v_n to $\Re^N$, let $v_n \rightharpoonup v$. One shows, using Lions' technique, that $v \neq 0$, and in fact, $v_n \to v$. Since then $|v|_p = 1$ and $\int_{\Re_+^N} |\nabla v|^2 = \hat{S}$ v is extremal and the Lemma follows.

We wish to end this note, pointing out how Theorem 1.3 fits in a more general framework. Namely, assuming $S_\lambda = \hat{S}$, (1.1) can be solved provided Γ_1 has a neighborhood $N_\delta(\Gamma_1)$, not homotopically equivalent to Γ_1, which can be sent via a continuous extension of $\sigma \to V_{\epsilon,\sigma}$, $\sigma \in \Gamma_1$, below the energy level S. More precisely, let us denote by $cat_{N_\delta(\Gamma_1)}(N_\delta(\Gamma_1), \Gamma_1)$ the category, in $N_\delta(\Gamma_1)$, of $N_\delta(\Gamma_1)$ relative to Γ_1 (see [6] for a definition of relative category and related variational principles). We have the following

Theorem 2.2 *Let $p = 2^*$, $S_\lambda = \hat{S}$. Assume*

$\exists \epsilon, \delta > 0$ *and* $\Phi \in C(N_\delta(\Gamma_1), \{E_\lambda \leq S - \epsilon\})$ *such that* $\Phi(\Gamma_1) \subset \{E_\lambda \leq \hat{S} + \epsilon\}$.

Then (1.1) has at least $cat_{N_\delta(\Gamma_1)}(N_\delta(\Gamma_1), \Gamma_1)$ non constant solutions, provided λ is large enough.

References

[1] **Adimurthi; G.Mancini:** The Neumann problem for elliptic equations with critical nonlinearity; in *Nonlinear Analysis*, a tribute in honour of Giovanni Prodi, Quaderni della Scuola Normale Superiore, Pisa, (1991), 9-25.

[2] **Adimurthi; G.Mancini:** Geometry and topology of the boundary in the critical Neumann problem; submitted to Crelles Journal.

[3] **Adimurthi; F.Pacella; S.L.Yadava:** Interaction between the geometry of the boundary and positive solutions of a semilinear Neumann problem with critical nonlinearity; submitted at Journal of Functional Analysis.

[4] **V.Benci; G.Cerami:** The effect of the domain topology on the number of positive solutions of non-linear elliptic problems, Arch. Rat. Mech. Anal. 114, (1991), 79-93.

[5] **V.Benci; G.Cerami; D.Passaseo**: On the number of the positive solutions of some nonlinear elliptic problems, preprint

[6] **G. Cerami; D.Passaseo**: Existence and multiplicity of positive solutions for nonlinear elliptic problems in exterior domains with "rich' topology, Nonlinear Analysis, T.M.A. vol 18, (1992), 109-119.

[7] **H.Brezis; L.Nirenberg**: Positive solutions of nonlinear elliptic equations involving critical exponents, Comm. Pure Appl. Math., vol. 36, 437-477 (1983).

[8] **P.L.Lions**: The concentration-compactness principle in the calculus of variations :the locally compact case. Part I,II, Ann.Inst¿ Poincaré Anal. Non Lineaire 1,109-145, and 223-283 (1984).

[9] **P.L.Lions**: The concentration compactness principle in the calculus of varations; Part.I, Revista Mathematica Iberoamericana, n. 1.1., (1985), 145-201.

[10] **G.Mancini; R.Musina**: The role of the boundary in some semilinear Neumann problems; to appear in Rend. Sem. Mat. Padova

[11] **W.M.Ni; I.Takagi**: On the shape of least- energy solutions to a semilinear Neumann problem, Comm. Pure Appl. Math. XLIV (1991) 819-851.

[12] **Xu-Jia Wang**: Neumann problems of semilinear elliptic equations involving critical Sobolev exponents; Journal of differential equations, (1991), 283-310.

[13] **Zhi Qiang Wang**: On the existence of multiple, single-peaked solutions for a semilinear Neumann problem; preprint.

[14] **Zhi Qiang Wang**: The effect of the domain geometry on the number of positive solutions of Neumann problems with critical exponents, preprint.

CLOSED ORBITS WITH PRESCRIBED ENERGY FOR SINGULAR CONSERVATIVE SYSTEMS WITH STRONGLY ATTRACTIVE POTENTIAL

LORENZO PISANI

Scuola Normale Superiore, Piazza dei Cavalieri 7, 56126 Pisa, Italy

§1. INTRODUCTION

The problem of finding periodic solutions of a natural conservative system

$$\ddot{q}(t) + \nabla V(q(t)) = 0, \tag{1.1}$$

either with prescribed period or with prescribed energy, is a classical one. In the last years a large amount of papers has been concerned with the case when V is a Keplerian-like potential, *i.e.* it is defined in $\mathbf{R}^N \backslash \{0\}$ and diverges for $x \to 0$ (most of references about these problems can be found in [2] or [11]).

In this survey paper we present some existence results for the fixed energy problem for the class of singular, strongly attractive potentials, *i.e.* for potentials such that

$$\limsup_{x \to 0} |x|^\alpha V(x) < 0 \tag{1.2}$$

for some $\alpha \geq 2$ (potentials which satisfy (1.2) with $0 < \alpha < 2$ are called weakly attractive). Let us remark that strongly attractive potentials satisfy the well known Strong Force condition introduced by Gordon (cf [9]).

As a simple model for this kind of potentials we can consider

$$V(x) = -\frac{a}{|x|^\alpha},$$

with $\alpha \geq 2$ and $a > 0$. In this case all periodic solutions of (1.1) have the form (see [8])

$$q(t) = r\left[\left(\cos\frac{2\pi t}{T}\right)\xi + \left(\sin\frac{2\pi t}{T}\right)\eta\right]$$

where $\xi, \eta \in S^{N-1}$, $\xi \cdot \eta = 0$ and r, T are strictly positive numbers such that

$$\left(\frac{2\pi}{T}\right)^2 = \frac{a\alpha}{r^{\alpha+2}}.$$

The corresponding energy is

$$\frac{1}{2}|\dot{q}(t)|^2 + V(q(t)) = \left(\frac{\alpha-2}{2}\right)\frac{a}{r^\alpha}.$$

Therefore, for $\alpha = 2$ the unique natural value of energy is 0, while, for $\alpha > 2$ we can choose every strictly positive value.

More in general, given a potential V, the admissible values of energy for periodic solutions of (1.1) are in the range of the function

$$W(x) = V(x) + \frac{1}{2}\nabla V(x) \cdot x.$$

Keeping in mind this facts, we are stating several results. Each one of them fits for a different class of potentials, which inherit some features of the model potentials (other results could be obtained by mixing the assumptions and the techniques used in the proofs). What is fixed in these theorems, obtained by variational methods, is the behaviour of V near the origin (assumption (1.2) or slightly stronger).

The first result is concerned with even potentials.

THEOREM 1.1. (Ambrosetti, cf [1]) *Let V satisfy*

(V0) $\qquad\qquad\qquad V \in C^1(\mathbf{R}^N \setminus \{0\}, \mathbf{R});$

(V1) $\qquad\qquad\qquad V(x) \leq 0;$

(V2) *there exists $\alpha > 2$ such that*

$$\limsup_{x \to 0} |x|^\alpha V(x) < 0;$$

(V3) $$V(x) = V(-x).$$

Then for all $h > 0$ there exists a solution of (1.1) such that

$$\frac{1}{2}|\dot{q}(t)|^2 + V(q(t)) = h. \tag{1.3}$$

The Theorem 1.1 covers a wide class of potentials, such as

$$V(x) = -\frac{1}{|x|} - \frac{A}{|x|^3},$$

which arises studying the motion of a particle under the attraction of a solid body (the coefficient A vanishes when the body is spherical). However it cannot be enlarged to the limit case of strongly attractive potentials, i.e. potentials such that

$$\limsup_{x \to 0} |x|^2 V(x) < 0. \tag{1.4}$$

For this kind of potential we are stating the following result, which improves the result [10].

THEOREM 1.2. Let V satisfy (V0-1) and
(V2′) there exist $\bar{h} > 0$ and $r > 0$ such that, for $0 < |x| \leq r$

$$V(x) + \frac{1}{2}\nabla V(x) \cdot x \geq \bar{h} \quad \text{and} \quad V(x) < 0;$$

(V4) $$\lim_{|x| \to \infty} \nabla V(x) = 0.$$

Then, for all $h \in]0, \bar{h}[$, there exists a solution of (1.1)-(1.3), such that $\|q\|_\infty > r$.

Let us remark that (V2′) is a bit stronger then (1.4) but it implies that V is not homogeneous of degree -2 near the origin. On the other hand, since our proof is based on the fact that $h > 0$, we have to exclude the homogeneous of degree -2 potentials, whose unique natural value of energy is 0.

As a corollary of Theorem 1.2 we obtain the following result.

THEOREM 1.3. *Let* V *satisfy (V0-1), (V4) and*

(V2″) *there exists* $\alpha > 2$ *and* $r > 0$ *such that, for* $0 < |x| \leq r$

$$\nabla V(x) \cdot x \geq -\alpha V(x) > 0.$$

Then, for all $h > 0$, *there exists a periodic solution of (1.1)-(1.3).*

The assumption (V2″) is clearly stronger than (V2) and, as a byproduct, it implies also (V2′); nevertheless let us remark that there is a kind of duality between Theorems 1.1 and 1.3: keeping fixed (V2″), the symmetry assumption (V3) can be replaced by the asymptotical assumption (V4).

The last result is different in nature. It was originally stated for regular potential defined in an annulus but it fits perfectly also for strongly attractive, Keplerian-like potentials.

THEOREM 1.4 (Benci-Giannoni, cf [5]) *Let* V *satisfy (V0). Suppose that there exist* $0 < r < R$ *and* $h \in \mathbf{R}$ *such that*

i) $\qquad\qquad\qquad V(x) < h \quad \text{for} \quad r \leq |x| \leq R;$

ii) $\qquad\qquad V(x) + \frac{1}{2}\nabla V(x) \cdot x \geq h \quad \text{for} \quad |x| = r;$

iii) $\qquad\qquad V(x) + \frac{1}{2}\nabla V(x) \cdot x \leq h \quad \text{for} \quad |x| = R.$

Then there exists a solution of (1.1)-(1.3) such that

$$r \leq |q(t)| \leq R.$$

Easy calculations show that the assumptions of Theorem 1.4 are fulfilled by the model potentials for $\alpha \geq 2$. Then, comparing Theorem 1.4 with the other results, we notice that *i)* is weaker than $(V0)$, *ii)* is weaker than $(V2')$, while *iii)* is essentially an asymptotical assumption; on the other hand this result makes clear where the solution is located.

§2. THE VARIATIONAL SETTING

The problem of finding periodic solutions of (1.1)-(1.3) is variational in nature. An useful variational principle is well known as Maupertuis-Jacobi Principle. It consists in finding the stationary points of the functional

$$\mathcal{F}(u) = \int_0^1 \sqrt{h - V(u)} |\dot{u}| \, dt \, ;$$

then the periodic solution of (1.1)-(1.3) are found after a suitable rescaling (see [4]).

We shall use a similar, slightly unusual variational principle, used also in [12] or [6] in order to find brake orbits for equation (1.1). Let

$$H := H^1(S^1, \mathbf{R}^N);$$

in H we consider the open set of non collision orbit

$$\Lambda := \{ u \in H \mid u(t) \neq 0, \, \forall t \}.$$

On Λ we can define the following C^1 functional

$$f(u) := \frac{1}{2} \int_0^1 |\dot{u}|^2 \, dt \cdot \int_0^1 (h - V(u)) \, dt \, ,$$

then the following theorem holds.

THEOREM 2.1. (cf e.g. [2]) Let $u \in \Lambda$ be a critical point of f such that $f(u) > 0$. Let

$$\omega^2 = \frac{\int_0^1 (h - V(u)) \, dt}{\frac{1}{2} \int_0^1 |\dot{u}|^2 \, dt},$$

Then $q(t) := u(\omega t)$ is a periodic solution of (1.1)-(1.3).

Before pointing out the main features of the funtional f, let us recall that in [3] and some other papers a slightly different variational principle has been used. First we notice that if $u \in \Lambda$ is a critical point of f at strictly positive level, then

$$\int_0^1 V(u) + \frac{1}{2} \nabla V(u) \cdot u \, dt = h.$$

Thus we consider the manifold

$$M_h = \{u \in \Lambda | \int_0^1 V(u) + \frac{1}{2} \nabla V(u) \cdot u \, dt = h\}$$

and for all $u \in M_h$, we have

$$f(u) = \mathcal{G}(u) := \frac{1}{4} \int_0^1 |\dot{u}|^2 \, dt \cdot \int_0^1 \nabla V(u) \cdot u \, dt.$$

In the quoted paper it has been shown that the critical points of $\mathcal{G}$ constrained on M_h are also free critical points of f. This variational principle, useful in the case of weakly attractive potentials, has the drawback of requiring $V \in C^2(\mathbf{R}^N \setminus \{0\}, \mathbf{R})$ and some technical assumptions on the second derivative of V.

Turning back to our funtional, we notice that from $(V1)$ and from our choice of $h > 0$ it follows that:

$$f(u) \geq \frac{h}{2} \int_0^1 |\dot{u}|^2 \, dt \geq 0 \tag{2.1}$$

Then the functional is bounded from below, but the minimum is achieved on the constants, at level 0. So, in order to apply the Theorem 2.1, we need a minimax procedure.

Unfortunately a straight application of classical minimax theorems does not seem possible because Λ is open and f does not satisfy the (PS) condition. Indeed the set

$$\{u \in \Lambda \,|\, f(u) = 0 \text{ and } f'(u) = 0\} = \mathbf{R}^N \setminus \{0\}$$

is non bounded and its closure, with respect to H, is not contained in Λ (we notice that this feature does not depend on V but depends only on the definition of f). Furthermore, as to strictly positive levels, we cannot exclude that a (PS) sequence converges to $0 \in \partial \Lambda$.

Finally let us remark that for potentials which are homogeneous of degree -2 a further degeneration occurs. We have already pointed out that

§3. SKETCHES OF THE PROOFS

PROOF OF THEOREM 1.1. The eveness of V allows us to avoid the lack of compactness at level 0. Indeed, if we consider the subspace of antiperiodic orbit

$$H_0 = \{u \in H \mid u(t + \frac{1}{2}) = -u(t)\},$$

any critical point u of f on $\Lambda_0 := \Lambda \cap H_0$ satisfies also $f'(u) = 0$. Furthermore, since $\|\dot{u}\|_2$ is a norm on H_0 (which is equivalent to the usual norm of H), by (2.1), the functional f is coercive on Λ_0.

As is usual in this setting, the behaviour of f at the boundary of Λ_0 is controlled by the growth assumption (V2); more precisely it implies that for any sequence $\{u_n\} \subset \Lambda_0$, if $u_n \rightharpoonup u \in \partial \Lambda_0$, then $\int_0^1 V(u_n)\,dt \to -\infty$. This property applies to our functional and implies that $f(u_n) \to \infty$; since f is a product, here the choice of $\alpha > 2$ plays its rôle.

Then, standard arguments yield the existence of an absolute minimum of f in Λ_0 and, since Λ_0 does not contain the constants, it is at a strictly positive level. This concludes the proof. □

PROOF OF THEOREM 1.2. Here we outline the proof of the Theorem 1.1 of [10], which is improved by the present result. A complete proof of Theorem 1.2 will be given in a forthcoming paper.

For every $\varepsilon > 0$, we introduce

$$B_\varepsilon = \{u \in H \mid \|u\| < \varepsilon\}$$

$$\Lambda_\varepsilon = \Lambda \setminus B_\varepsilon$$

STEP 1. For every $\varepsilon > 0$, the functional f restricted to Λ_ε satisfies (PS) at strictly positive levels.

STEP 2. For every $\varepsilon > 0$, there exists $a > 0$ such that $\{x \in \mathbf{R}^N \mid |x| \geq \varepsilon\}$ is a deformation retract of $f^a \cap \Lambda_\varepsilon$ in Λ_ε, where

$$f^a = \{u \in \Lambda \mid f(u) \leq a\}.$$

STEP 3. There exists $\bar{\varepsilon} > 0$ such that $\Lambda_{\bar{\varepsilon}}$ is positively invariant for the steepest descend flow defined by
$$\begin{cases} \frac{\partial}{\partial s}\eta(s,u) = -\nabla f(\eta(s,u)) \\ \eta(0,u) = u. \end{cases}$$
In order to define this flow, in [10] it was required that ∇V was locally lipschitz continuous; furthermore, in order to have a global semiflow the following asymptotical assumption was required

(V5) $$\liminf_{|x|\to\infty} V(x) > -\infty.$$

Since $\Lambda_{\bar{\varepsilon}}$ is a deformation retract of Λ it has same rich topology, then we can apply the Lusternik-Schnirelman theory. For all $n \in \mathbf{N}$, let
$$\mathcal{K}_n = \{X \subset \Lambda_{\bar{\varepsilon}} \mid X \text{ is compact and } \mathrm{cat}_{\Lambda_{\varepsilon}}(X) \geq n\}$$
$$c_n = \inf_{X \in \mathcal{K}_n} \max_{u \in X} f(u).$$
It is known (see [7]) that $\mathcal{K}_n \neq \emptyset$ for all $n \in \mathbf{N}$; furthermore, since f is non negative, $c_n \geq 0$. By Step 2 we obtain that $c_3 > 0$; then standard arguments of LS theory, developed in a set which is invariant under the steepest descent flow, imply that c_3 is a critical value of f.

Finally the L^∞ estimate on the solution is proved arguing by contradiction; indeed if $\|u\|_\infty \leq r$ then, by $(V2'')$, we get
$$\bar{h} > h = \int_0^1 V(u) + \frac{1}{2}\nabla V(u) \cdot u \, dt \geq \bar{h}.$$

The improvement is concerned with Step 3: under the hypotheses of Theorem 1.2 we can find Ψ pseudogradient vector field for f and $\bar{\varepsilon} > 0$ such that $\Lambda_{\bar{\varepsilon}}$ is positively invariant under the global semiflow defined by
$$\Phi(u) = \begin{cases} -\chi(u)\Psi(u)/\|\Psi(u)\| & \text{if } f(u) > 0 \\ 0 & \text{otherwise,} \end{cases}$$
χ being a suitable cut-off function. Let us remark that as critical value we can choose also
$$c = \inf_{X \in \mathcal{A}} \max_{x \in X} f(x),$$
where the class of admissible sets is defined as follows
$$\mathcal{A} = \{X \subset \Lambda_{\bar{\varepsilon}} \mid X \text{ is compact and}$$
$$\text{cannot be retracted on } \mathbf{R}^N \text{ without crossing } \partial\Lambda_{\bar{\varepsilon}}\}$$
$\square$

PROOF OF THEOREM 1.3. First we recall that from $(V2'')$ it follows that
$$V(x) \leq -\frac{c}{|x|^\alpha}, \qquad \forall x \in \mathbf{R}^N \setminus \{0\} \text{ with } |x| \leq r,$$
being c a suitable positive constant. Again from $(V2'')$ it follows
$$V(x) + \frac{1}{2} \nabla V(x) \cdot x \geq -\frac{\alpha - 2}{2} V(x) \geq \frac{(\alpha - 2)}{2} \frac{c}{|x|^\alpha}.$$

Then for every $h > 0$ we can find $r' \in {]}0, r]$ such that V satisfies $(V2')$ with $\bar{h} = h + 1$. Thus Theorem 1.2 applies and Theorem 1.3 follows. □

PROOF OF THEOREM 1.4. Consider in H the open subset
$$\Lambda^* := \{u \in H \mid r < |q(t)| < R\} \subset \Lambda.$$

Let $\chi \in C^2({]}r, R[, \mathbf{R}_+)$ be such that:

$$\chi(s) = \frac{1}{(s - r)^2} \qquad \text{if } s \text{ is "near" } r, \tag{4.1}$$

$$\chi(s) = \frac{1}{(s - R)^2} \qquad \text{if } s \text{ is "near" } R, \tag{4.2}$$

$$\chi(s) = 0 \qquad \text{if } s \text{ is "near" } \frac{r + R}{2}, \tag{4.3}$$

$$\chi''(s) \geq 0 \qquad \text{for every } s \in {]}r, R[. \tag{4.4}$$

Now consider the family $\{f_\varepsilon\} \subset C^1(\Lambda^*, \mathbf{R})$ of penalized functionals
$$f_\varepsilon(u) := f(u) + \varepsilon \int_0^1 \chi(|u|) \, dt \qquad \varepsilon > 0.$$

For every $\varepsilon > 0$, the functional f_ε satisfies the (PS) condition; then, using the method of Lusternik and Fet to find non trivial closed geodesics, we can find u_ε, a non constant critical point of f_ε such that $f_\varepsilon(u_\varepsilon) \geq const > 0$.

The next step is "pass to the limit" for $\varepsilon \to 0$. There exists a sequence $\varepsilon_n \to 0$ such that $u_{\varepsilon_n} \to u \in \overline{\Lambda^*}$, weakly in $H^2(S^1, \mathbf{R}^N)$. The limit u satisfies the following equation

$$\left(\int_0^1 (h - V(u)) d\tau \right) \ddot{u}(t) + \left(\frac{1}{2} \int_0^1 |\dot{u}|^2 d\tau \right) \nabla V(u(t)) = \lambda(t) \frac{u(t)}{|u(t)|} \tag{4.5}$$

for a.e. $t \in [0,1]$ where λ is a function of $L^2(0,1;\mathbf{R})$ such that

$$\lambda(t) = 0 \quad \text{for every } t \in [0,1] \text{ such that } r < |u(t)| < R, \qquad (4.6)$$

$$\lambda(t) \leq 0 \quad \text{for a.e. } t \in [0,1] \text{ such that } |u(t)| = r, \qquad (4.7)$$

$$\lambda(t) \geq 0 \quad \text{for a.e. } t \in [0,1] \text{ such that } |u(t)| = R. \qquad (4.8)$$

The assumptions *ii)* and *iii)* allow us to conclude $\lambda(t) = 0$ a.e. in $[0,1]$. Sobstituting in (4.5) we get that u is a critical point of f such that $r \leq |u(t)| \leq R$. By standard arguments, the value $f(u)$ is strictly positive and this completes the proof. □

REFERENCES

[1] A. Ambrosetti, *Nonlinear analysis, a tribute in honour of Giovanni Prodi*, Quaderni Scuola Norm. Sup. Pisa (1991), 51-60.

[2] A. Ambrosetti, *Supplément au Bulletin de la Société Mathématique de France*, Mémoire N° 49, **120** (1992).

[3] A. Ambrosetti, V. Coti Zelati, *Archive Rat. Mech. Analysis* **112** (1990), 339-362.

[4] V. Benci, *Ann. Inst. H. Poincaré, Analyse Nonlineaire* **1** (1984), 401-412.

[5] V. Benci, F. Giannoni, *J. Diff. Eq.* **82** (1989), 60-70.

[6] V. Benci, F. Giannoni, *Advanced Topics in the Theory of Dynamical Systems*, G. Fusco, M. Iannelli, L. Salvadori eds., Academic Press (1989), 37-49.

[7] E. Fadell, S. Husseini, *Proc. Amer. Math. Soc.* **107** (1989), 527-536.

[8] F. Giannoni, *Boll. Un. Mat. Ital.* (7) **3 - B** (1989), 547-577.

[9] W.B. Gordon, *Trans. Amer. Math. Soc.* **204** (1975), 113-135.

[10] L. Pisani, *J. Nonlinear Analysis*, to appear.

[11] S. Terracini, *Ph.D. Thesis*, SISSA (1990).

[12] E.W.C. Van Groesen, *J. Math. Anal. Appl.* **132** (1988), 1-12.

ON A MODEL PROBLEM OF MULTIPLE PENDULUM TYPE

PAUL H. RABINOWITZ

Department of Mathematics and Center
for the Mathematical Sciences
University of Wisconsin
Madison, Wisconsin, 53706

The goal of this note is to describe some recent work on the existence of heteroclinic and homoclinic orbits for a model equation of multiple pendulum type. By way of motivation, consider the Hamiltonian system

$$\ddot{q} + V'(q) = 0. \tag{1.1}$$

Here $q \in \mathbf{R}^n$ and V satisfies

(V) $V \in C^1(\mathbf{R}^n, \mathbf{R})$ and is T_i periodic in q_i, $1 \leq i \leq n$.

A Hamiltonian system satisfying (V) will be said to be of *multiple-pendulum type*. For the actual multiple pendulum system of equations, (and $n \geq 2$) the $\ddot{q}$ term is replaced by a quasilinear one. Let

$$\mathcal{M} = \{x \in \mathbf{R}^n \mid V(x) = \max_{\xi \in \mathbf{R}^n} V(\xi)\}.$$

Assume

($\mathcal{M}$) consists of isolated points.

This will be true for generic V. If

$$\lim_{t \to \infty} x(t) = x_0,$$

we write $x(\infty) = x_0$. In [1], it was proved that under hypotheses (V) and ($\mathcal{M}$), (1.1) possesses heteroclinic orbits joining points in $\mathcal{M}$. More precisely

---

This research was sponsored in part by the US Army Research Office under contract #DAAL03-87-K-0043 and by the National Science Foundation under Grant #MCS-8110556. Any reproduction for the purpose of the United States Government is permitted.

Theorem 1.2
If (V) and ($\mathcal{M}$) are satisfied, for each $\xi \in \mathcal{M}$, there is an $\eta \in \mathcal{M}\setminus\{\xi\}$ and a heteroclinic orbit q of (1.1) joining ξ and η, i.e. $q(-\infty) = \xi$, $q(\infty) = \eta$, and $\dot{q}(-\infty) = 0 = \dot{q}(\infty)$.

Since (1.1) is time reversible, it follows that there also is a heteroclinic orbit of (1.1) joining η and ξ. Theorem 1.2 is proved by an elementary minimization argument. In [2], a somewhat more elaborate minimization argument was used to obtain a more general result:

Theorem 1.3
If (V) and ($\mathcal{M}$) are satisfied, for each ξ and $\eta \in \mathcal{M}$ with $\xi \neq \eta$, there exists a heteroclinic chain of solutions of (1.1) joining ξ and η.

By a *heteroclinic (k−) chain* joining ξ and η is meant a collection of $k+1$ points $\xi_0 = \xi, \xi_1, \ldots, \xi_k = \eta$ with $\xi_i \in \mathcal{M}$, $i = 0, \ldots, k$, such that $\xi_i \neq \xi_j$ for $i \neq j$, and corresponding heteroclinic orbits q_i of (1.1) joining ξ_i to ξ_{i+1}, $0 \leq i \leq k-1$. We say a solution of (1.1) is of *k-chain type* if it is "near" a heteroclinic k-chain.

As a simple example of (1.1) under (V) and ($\mathcal{M}$), suppose $n = 1$ and $V(x) = \cos x - 1$. Then (1.1) is the classical simple pendulum equation. For this case the phase plane shows there are two heteroclinic orbits emanating from and two terminating at each $\xi \in \mathcal{M}$. Some multiplicity results for (1.1) under (V) and ($\mathcal{M}$) which contain the simple pendulum as a special case have been obtained in [1] and [3]. Likewise analogues of Theorem 1.2 for the actual multiple pendulum were given by Felmer [4].

Our goal now is to study (1.1) when the potential energy V also depends on t in a time periodic fashion. To simplify matters, however, V will be chosen for which the analogue of $\mathcal{M}$ contains only two points. Thus consider:

(HS) $\ddot{q} + V_q(t,q) = 0$

where V satisfies

(V_1) $V \in C^2(\mathbf{R} \times \mathbf{R}^n, \mathbf{R})$ and is T periodic in t,

(V_2) there is a $\xi \neq 0$ such that for all $t \in [0,T]$, $V(t,0) = 0 = V(t,\xi)$, $V_q(t,0) = 0 = V_q(t,\xi)$, and $L_0(t) \equiv V_{qq}(t,0)$, $L_1(\xi) \equiv V_{qq}(t,\xi)$ are negative definite,

(V_3) $V(t,q) < 0$ if $q \neq 0, \xi$,

(V_4) there is a $V_0 < 0$ such that $\varlimsup_{|q| \to \infty} V(t,q) \leq V_0$.

The purpose of (V_4) is to avoid problems as $|q| \to \infty$ and it can be weakened.

Note that 0 and ξ are equilibrium solutions of (HS) (and global maxima of $V(t, \cdot)$). Moreover as for (1.1) we have:

Proposition 1.3
If V satisfies (V_1)–(V_4), there are heteroclinic solutions v and w of (HS) such that $v(-\infty) = 0 = w(\infty)$, $v(\infty) = \xi = w(-\infty)$, $\dot{v}(\pm\infty) = 0 = \dot{w}(\pm\infty)$.

Indeed, let

$$\Gamma(0,\xi) = \{q \in W^{1,2}_{\text{loc}}(\mathbf{R},\mathbf{R}^n) \mid q(-\infty) = 0, q(\infty) = \xi \quad (1.4)$$

and $\int_{-\infty}^{\infty}|\dot{q}|^2 dt < \infty\}$,

$$\Gamma(\xi,0) = \{q \in W^{1,2}_{\text{loc}}(\mathbf{R},\mathbf{R}^n) \mid q(-\infty) = \xi, q(\infty) = 0 \qquad (1.5)$$

and $\int_{-\infty}^{\infty}|\dot{q}|^2 dt < \infty\}$,

and

$$I(q) = \int_{-\infty}^{\infty}\left(\frac{1}{2}|\dot{q}|^2 - V(t,q)\right) dt. \qquad (1.6)$$

Define

$$c(0,\xi) = \inf_{q \in \Gamma(0,\xi)} I(q) \qquad (1.7)$$

$$c(\xi,0) = \inf_{q \in \Gamma(\xi,0)} I(q). \qquad (1.8)$$

Given this functional framework, variants of arguments from [1] easily establish the existence of $v \in \Gamma(0,\xi)$ and $w \in \Gamma(\xi,0)$ such that $I(v) = c(0,\xi)$ and $I(w) = c(\xi,0)$.

Observe that if $q \in \Gamma(0,\xi) \cup \Gamma(\xi,0)$, $j \in \mathbf{Z}$, and

$$\tau_j q(t) = q(t - jT), \qquad (1.9)$$

then by (V_1),

$$I(\tau_j q) = I(q). \qquad (1.10)$$

Thus the functions v and w are not unique minimizers of I in their respective classes of translates. Normalize v and w in any convenient way to make them unique. E.g. let $B_r \equiv \{x \in \mathbf{R}^n \mid |x| < r\}$. Let $\epsilon_1 \ll 1$ and choose v so that $v(t) \in \overline{B}_{\epsilon_1}$ for $t \leq s_1$ and v exits from $\overline{B}_{\epsilon_1}$ for the first time at $t = s_1$ for some $s_1 \in [0,T)$.

Consider $J(\varphi) = I(v+\varphi)$ where $\varphi \in W^{1,2}(\mathbf{R},\mathbf{R}^n)$. It is straightforward to verify that $J \in C^1(W^{1,2}(\mathbf{R},\mathbf{R}^n),\mathbf{R})$ and 0 is a critical point of J. See e.g. [5] for a related computation. Henceforth we *assume* that v and w are isolated critical points of I in the sense that there is a $\nu > 0$ such that if $E = W^{1,2}(\mathbf{R},\mathbf{R}^n)$ with norm $\|\cdot\|$, $\varphi \in E\setminus\{0\}$, and $I'(v+\varphi) = 0$ or $I'(w+\varphi) = 0$, then $\|\varphi\| \geq \nu$. Since v and w are minima of I in $\Gamma(0,\xi)$ and $\Gamma(\xi,0)$ respectively, this seems to be a reasonable assumption. However it is difficult to verify in explicit examples.

Our goal now is to prove

<u>Theorem 1.11</u>
Suppose V satisfies (V_1)–(V_4) and v and w are isolated critical points of I. Then (HS) possesses infinitely many solutions of k-chain type for each $k \in \mathbf{N}$, $k \geq 2$.

A sketch of the proof of Theorem 1.11 will be given for $k = 2$. See [6] for more details. The proof relies on ideas from Coti Zelati - Ekeland - Séré [7], Séré [8], Giannoni - Rabinowitz [9], and [5] as well as some new ones. The 2-chain solutions (and k-chain solutions for even k) will lie in E. Arguments from [5] show $I \in C^1(E, \mathbf{R})$. Define

$$\Gamma_2 = \{q \in E \mid \text{ there is a } p \in \mathbf{R} \text{ such} \tag{1.12}$$
$$\text{that } q \equiv \xi \text{ near } p\}.$$

Solutions of 2-chain type will be obtained near Γ_2. Set

$$b_2 = \inf_{q \in \Gamma_2} I(q). \tag{1.13}$$

The following simple proposition is a key tool in the proof of Theorem 1.11.

Proposition 1.14
$b_2 = c(0, \xi) + c(\xi, 0)$.

Proof.
If $q \in \Gamma_2$, set $q^- = q|_{(-\infty, p]}$ and $q^+ = q|_{[p, +\infty)}$. Then q^-, q^+ extend to $Q^- \in \Gamma(0, \xi)$ and $Q^+ \in \Gamma(\xi, 0)$ (e.g. via $Q^-(t) = q^-(t), t \in (-\infty, p], Q^-(t) = \xi, t > p$, etc.). Moreover

$$I(Q^-) = \int_{-\infty}^{p} \left(\frac{1}{2} |\dot{q}^-|^2 - V(t, q^-) \right) dt \tag{1.15}$$

and

$$I(Q^+) = \int_{p}^{\infty} \left(\frac{1}{2} |\dot{q}^+|^2 - V(t, q^+) \right) dt. \tag{1.16}$$

Thus (1.12)–(1.13) and (1.15)–(1.16) readily show that $b_2 \geq c(0, \xi) + c(\xi, 0)$. On the other hand, for any $\epsilon > 0$, appropriately truncating v and a translate of w yields $q_\epsilon \in \Gamma_2$ such that

$$I(q_\epsilon) \leq c(0, \xi) + c(\xi, 0) + \epsilon$$

and hence the Proposition.

Now, roughly speaking, the strategy of the proof of Theorem 1.11 is to show that if (HS) does not possess infinitely many solutions of 2-chain type, then there is a $Q \in \Gamma_2$ such that

$$I(Q) < b_2 \tag{1.17}$$

contrary to Proposition 1.14.
To continue, a family of approximations to the formal sum of v and translates $\tau_\ell w$ of v with ℓ large is needed. Let $\rho_1 < |\xi|/3$. For $\rho \leq \rho_1$ and

$u \in \{v, w,$ and their translates$\}$, define $\alpha(\rho, u)$ to be the unique maximal $t \in \mathbf{R}$ such that $u(t) \in \overline{B}_\rho(u(-\infty))$ for all $t \leq \alpha(\rho, u)$ and $\omega(\rho, u)$ as the unique minimal $t \in \mathbf{R}$ such that $u(t) \in \overline{B}_\rho(u(\infty))$ for all $t \geq \omega(\rho, u)$. Note that $\alpha(\rho, \tau_j u) = \alpha(\rho, u) + jT$ for all $j \in \mathbf{Z}$ and similarly for ω. Moreover by (V_1)-(V_2) and (HS), $\alpha(\rho, u) \to -\infty$ and $\omega(\rho, u) \to \infty$ as $\rho \to 0$. Now a family of approximate two chains $\mathcal{P}(\rho, \ell)$ can be defined as follows

$$\begin{aligned}
\mathcal{P}(\rho, \ell) = 0 &= v(-\infty),\ t \in (-\infty, \alpha(\rho, v) - 1), \qquad (1.18)\\
&= [t - \alpha(\rho, v) + 1)v(\alpha(\rho, v))\\
&\quad \text{i.e. a convex combination of } 0 \text{ and } v(\alpha(\rho, v)),\\
&\qquad t \in [\alpha(\rho, v) - 1, \alpha(\rho, v)],\\
&= v(t),\quad t \in [\alpha(\rho, v), \omega(\rho, v)],\\
&= \text{convex combination of } v(\omega(\rho, v)) \text{ and } \xi,\\
&\qquad t \in [\omega(\rho, v), \omega(\rho, v) + 1],\\
&= \xi = v(\infty) = w(-\infty),\ t \in [\omega(\rho, v) + 1, \alpha(\rho, \tau_\ell w) - 1],\\
&= \text{convex combination of } \xi \text{ and } \tau_\ell w(\alpha(\rho, \tau_\ell w)),\\
&\qquad t \in [\alpha(\rho, \tau_\ell w) - 1, \alpha(\rho, \tau_\ell w)],\\
&= \tau_\ell w,\ t \in [\alpha(\rho, \tau_\ell w), \omega(\rho, \tau_\ell w)],\\
&= \text{convex combination of } \tau_\ell w(\rho, \tau_\ell w) \text{ and } 0,\\
&\qquad t \in [\omega(\rho, \tau_\ell w), \omega(\rho, \tau_\ell w) + 1],\\
&= 0 = w(\infty),\ t \in (\omega(\rho, \tau_\ell w) + 1, \infty).
\end{aligned}$$

Observe that in order for $\mathcal{P}(\rho, \ell)$ to be defined, it is necessary that $\alpha(\rho, \tau_\ell w) - \omega(\rho, v) \geq 2$ or $\ell T \geq 2 + \omega(\rho, v) - \alpha(\rho, w)$ which will be the case if ℓ/ρ is large compared to 1.

By its definition, $\mathcal{P}(\rho, \ell) \in \Gamma_2$. Moreover $I(\mathcal{P}(\rho, \ell)) \to b_2$ as $\rho \to 0$ uniformly in ℓ. Now suppose there are only a finite number (modulo the $\mathbf{Z}$ symmetry of (1.10)) of solutions of 2-chain type of (HS). Then $\mathcal{P}(\rho, \ell)$ cannot lie near this set of solutions for any sufficiently large $\ell \in \mathbf{N}$. Moreover we have

Proposition 1.19

There is an $r_0 > 0$ and functions ρ, ϵ, ℓ_0 on $\mathbf{R}^+$ and η on E defined for $0 < r \leq r_0$ such that whenever $0 < \rho \leq \rho_0(r)$ and $\ell \geq \ell_0(r)$, then $I(\mathcal{P}(\rho, \ell)) \leq b_2 + \epsilon(r)$ implies that

$$I(\eta(\mathcal{P}(\rho, \ell))) \leq b_2 - \epsilon(r) \qquad (1.20)$$

and

$$\|\eta(\mathcal{P}(\rho, \ell)) - \mathcal{P}(\rho, \ell)\| \leq r. \qquad (1.21)$$

Idea of proof

If I satisfied the Palais-Smale condition (PS) the result would be a standard one. Unfortunately (PS) does not hold here as is easy to see. Nevertheless following [7] or [5], it is not difficult to analyze the behavior of Palais-Smale sequences. Moreover lower bounds for $\|I'\|$ can be obtained in sets of the form $\mathcal{B}_r(\mathcal{P})\backslash\mathcal{B}_{r/8}(\mathcal{P})$ where $\mathcal{B}_r(u)$ denotes an open ball in E of radius r about u. These facts then yield the Proposition. See [6].

Remark 1.22
Proposition 1.19 leads in fact to a stronger result than Theorem 1.11. Let $\mathcal{K}$ denote the set of critical points of I e.g. in E. Suppose $r \leq r_0$ and $\rho \leq \rho_0(r)$. Then $\mathcal{B}_{2r}(\mathcal{P}(\rho,\ell)) \cap \mathcal{K} \neq \emptyset$ for all but finitely many $\ell \in \mathbf{N}$. In addition the corresponding critical values lie near b_2.

Continuing with the proof of Theorem 1.11, if $\mathcal{Q} \equiv \eta(\mathcal{P}(\rho,\ell))$ were in Γ_2, then (1.20) would be contrary to Proposition 1.14, thereby completing the proof. However in general, $\mathcal{Q} \notin \Gamma_2$; it is merely close to Γ_2 as expressed by (1.21). Thus $\mathcal{Q}$ must be modified to obtain an element $Q \in \Gamma_2$ satisfying (1.17). Towards that end, let $\rho \leq \rho_0(r)$, $\ell \geq \ell_0(r)$, $\overline{\omega} = \omega(\rho, v) + 1$, and $\overline{\alpha} = \alpha(\rho, \tau_\ell w) - 1$. Note that for fixed ρ, $\overline{\omega}$ and $\overline{\alpha}$ depend on ℓ. That ℓ can be chosen arbitrarily large is crucial for what follows.

Let

$$\tilde{E} = \{u \in W^{1,2}([\overline{\omega},\overline{\alpha}], \mathbf{R}^n) \mid u(\overline{\omega}) = \mathcal{Q}(\overline{\omega}), \quad (1.23)$$
$$u(\overline{\alpha}) = \mathcal{Q}(\overline{\alpha}), \text{ and}$$
$$\|u - \mathcal{P}(\rho,\ell)\|_{W^{1,2}([\overline{\omega},\overline{\alpha}],\mathbf{R}^n)} \leq 8r\}.$$

Observe that $\mathcal{Q}\big|_{[\overline{\omega},\overline{\alpha}]} \in \tilde{E}$ so $\tilde{E} \neq \emptyset$. Moreover by (1.21) and (1.23), $u \in \tilde{E}$ implies u is close to ξ in an L^∞ sense. Hence if

$$\Psi(u) = \int_{\overline{\omega}}^{\overline{\alpha}} \left(\frac{1}{2}|\dot{u}|^2 - V(t,u)\right) dt, \quad (1.24)$$

then by (V_2), Ψ is convex on $\tilde{E}$. Therefore the variational problem

$$\inf_{u \in \tilde{E}} \Psi(u) \quad (1.25)$$

possesses a unique solution $x \in \tilde{E}$. Thus defining

$$U(x) = x(t), \quad t \in [\overline{\omega},\overline{\alpha}], \quad (1.26)$$
$$= \mathcal{Q}(t), \quad t \notin [\overline{\omega},\overline{\alpha}],$$

then by the definition of Ψ,

$$I(U) \leq I(\mathcal{Q}) \leq b_2 - \epsilon. \quad (1.27)$$

Moreover the choice of $\tilde{E}$ and Ψ imply $x(t)$ is a classical solution of (HS) on $(\overline{\omega},\overline{\alpha})$ with $x(\overline{\omega}) = \mathcal{Q}(\overline{\omega})$, $x(\overline{\alpha}) = \mathcal{Q}(\overline{\alpha})$. Recall again that these facts are true for all $\ell \in \mathbf{N}$ sufficiently large.

Now an argument exploiting this fact will be used. Let

$$\mu = \frac{1}{2}(\overline{\omega} + \overline{\alpha}) = \frac{1}{2}(\omega(\rho, v) + \alpha(\rho, w) + \ell T). \tag{1.28}$$

A comparison function argument based on the Maximum Principle shows that there are constants γ, $A > 0$ and independent of $\ell \geq \ell_0$ such that

$$\|x(t) - \xi\|_{C^1([\mu-\frac{1}{2}, \mu+\frac{1}{2}], \mathbf{R}^n)} \leq Ae^{-\gamma \ell}. \tag{1.29}$$

Thus the function U can be modified in e.g. $[\mu - \frac{1}{4}, \mu + \frac{1}{4}]$ to produce $Q \in \Gamma_2$ which by (1.29) is very close to U for large ℓ and therefore (1.17) holds. This gives the desired contradiction and completes the proof of Theorem 1.11.

REFERENCES

1. P. H. Rabinowitz, Ann. Inst. H. Poincare–Analyse nonlineaire, 6, 331–346 (1989).

2. P. H. Rabinowitz, Frontiers in Pure and Applied Mathematics, (R. Dautray, ed.) 267–278 (1991).

3. P. H. Rabinowitz, to appear Proc. Conf. on Variational Methods in Hamiltonian Systems and Elliptic Equations.

4. P. Felmer, Ann. Inst. H. Poincare–Analyse nonlineaire, 8, 477–497 (1991).

5. V. Coti Zelati and P. H. Rabinowitz, J. Amer. Math. Soc., 4, 693–727 (1991).

6. P. H. Rabinowitz, to appear, Calculus of Variations and Partial Differential Equations.

7. V. Coti Zelati, I. Ekeland, and E. Séré, Math. Ann., 288, 133–160 (1990).

8. E. Séré, to appear Math. Z.

9. F. Giannoni and P. H. Rabinowitz, to appear.

PERTURBATION OF SYMMETRY AND GENERALIZED MOUNTAIN PASS THEOREMS

BERNHARD RUF

Dip. di Matematica, Università degli Studi, Via Saldini 50, Milano, Italy

1. MINIMAX PRINCIPLES AND INDEX-THEORIES

We recall the following minimax principle of *R. Palais* [P].

Let E denote a Banach space, and assume that $J : E \to R$ is a C^1-functional satisfying the Palais-Smale condition (a compactness condition, see e.g. [S1]). Furthermore assume that $\mathcal{A}$ is a class of subsets of E which is *invariant under the gradient flow* generated by the functional J, i.e. if $\eta : [0, +\infty) \times E \to E$ denotes this flow, then $\eta(t, A) \in \mathcal{A}$ for every $A \in \mathcal{A}, t \geq 0$. Then set

$$c = inf_{A \in \mathcal{A}} sup_{u \in A} J(u) .$$

One proves that if $c > -\infty$ then c is a critical value of J; see e.g. Struwe [S1].

As a classsical example for this principle we mention the Courant-Weyl variational principle for finding the eigenvalues of the Laplacian: Let $\Omega \subset R^n$ denote a bounded domain, and consider

$$-\Delta u = \lambda u \text{ in } \Omega , \ u = 0 \text{ on } \partial\Omega . \qquad (1)$$

To find the eigenvalues one takes the functional

$$J(u) = \int_\Omega |\nabla u|^2 dx \ : H_0^1(\Omega) \to R ;$$

the $k-th$ eigenvalue is then given by
$$\lambda_k = inf_{E_k \in \mathcal{E}_k} sup_{u \in E_k \cap S_{L^2}} J(u) ,$$

where S_{L^2} denotes the unit L^2- sphere in H_0^1 and $\mathcal{E}_k$ denotes the set of all $k-$dimensional, linear subspaces E_k of H_0^1. Here the invariance of the class $\mathcal{E}_k$ is due to the linearity of the gradient flow $\eta(t, \cdot)$.

In the case of a nonlinear equation the gradient flow of the associated functional will be nonlinear, and hence it proves more difficult to find invariant families $\mathcal{A}$. However, if the problem is invariant under certain group actions the above procedure can be repeated. The first to observe this was Krasnoselskii [K] (see also Coffman [C]) for the case that the problem is symmetric with respect to the group $Z_2 = \{\pm id\}$ (i.e. if u solves the equation, then also $-u$). Krasnoselskii's idea is to introduce a device which measures the topological complexity (generalized dimension) of symmetric (i.e. invariant under $\pm id$) subsets of E. The definition of the (equivalent) $Z_2 - genus$ of Coffman is the following:

Let $\mathcal{S} = \{A \subset E \setminus \{0\}; \ A \ closed, \ symmetric\}$ and for $A \in \mathcal{S}$ set
$$\gamma(A) = min\{n \in N \ ; \ \exists \ g : A \to R^n \setminus \{0\} \ , \ g \ continuous \ and \ odd\} .$$

The genus has (among others) the following properties:
1. (invariance) Let $A \in \mathcal{S}$, and $h : A \to E$ be continuous and odd. Then $\gamma(A) \leq \gamma(h(A))$.
2. (dimension) Let $E_k \subset E$ denote a $k-$dimensional subspace, let S be the unit sphere in E, and $S^{k-1} = S \cap E_k$. Then $\gamma(S^{k-1}) = k$.

The proof of 1. follows immediatly from the definition, the proof of 2. is a direct consequence of the $Borsuk - Ulam$ theorem.

Based on the genus one can now proceed as before to obtain critical points of functionals related to semilinear equations. For example, let $g : \overline{\Omega} \times R \to R$ be continuous and odd (and satisfying a suitable growth-condition), and consider the nonlinear eigenvalue problem

$$-\Delta u + g(x, u) = \lambda u \ in \ \Omega \ , \ u = 0 \ on \ \partial \Omega . \qquad (2)$$

Setting, for $\rho > 0$, $\mathcal{B}_k(\rho) = \{B \subset \rho S_{L^2} \subset H_0^1(\Omega) \ ; \gamma(B) \geq k\}$ one finds the eigenvalues of (2) by

$$\mu_k(\rho) = inf_{B \in \mathcal{B}_k} sup_{u \in B} J(u) ,$$

where
$$J(u) = \frac{1}{2} \int_\Omega |\nabla u|^2 + \int_\Omega G(x, u) , \qquad (3)$$

with $G(x,s) = \int_0^s g(x,t)dt$ a primitive of g.

Similar constructions can be done for other compact groups provided a Borsuk-Ulam theorem is available. Indeed, following the concept of Coffman, Benci [B] has introduced for the circle group an $S^1 - index$ (or $S^1 - genus$) as follows:

Suppose that E is a separable Banach space which carries an S^1-action $T_\theta, \theta \in [0, 2\pi)$. Let R_θ, $\theta \in S^1$, denote an S^1- action on C^k. R_θ is *regular*, if the fixed point set consists of the origin: $FixR_\theta = \{0\}$. Furthermore, a mapping $F : E \to C^k$ is *equivariant* with respect to (T_θ, R_θ), if $F \circ T_\theta = R_\theta \circ F$, for all $\theta \in S^1$. For a T_θ-invariant set $A \subset H$ (i.e. $T_\theta \subseteq A$, for all $\theta \in S^1$) one defines, for $k \in N$, the classes of maps

$$M_k(A, R) = \{\phi : A \to C^k \setminus \{0\}; \phi \text{ is equivariant w.r.t. } (T_\theta, R_\theta)\}.$$

For sets in the class $\mathcal{E} = \{A \subset H; A \cap FixT_\theta = \emptyset, A \text{ closed and } T_\theta\text{-invariant}\}$ one defines

$$i_{S^1}(A) = \inf\{k \in N; \exists \text{ regular } S^1 - \text{action } R_\theta \text{ on } C^k \text{ s. th. } M_k(A, R) \neq \emptyset\}.$$

For this index the same properties hold as for the Z_2-index. Note however that the dimension property now measures the *complex* (generalized) dimension of a set.

As an example we consider the problem of characterizing variationally the periodic *Fučik - spectrum* of $-d^2/dt^2$, i.e. the values μ and ν such that the equation

$$-u'' = \mu u^+ - \nu u^- , \quad u \; T-periodic, \tag{4}$$

has a non-trivial solution. These values can be found as critical points of the functional

$$J_\gamma(u) = \int_0^T |u'|^2 - \gamma|u^+|^2 : H_T^1 \to R, \tag{5}$$

with $\gamma = \mu - \nu$ fixed, and $u \in S_{L^2} \subset H_T^1$, the Sobolev-space of T-periodic functions. Clearly, the functional J_γ is not even, i.e. not Z_2-invariant; however, it is invariant with respect to the S^1-action given by the time shifts: $T_\theta u(t) = u(t + \theta)$. Note however that this action is not fixed point free: $FixT_\theta$ consists of the constant functions $E_0 \sim R$. An extension of the S^1-index to this situation is possible, see [BLMR], and hence we can again define classes $\mathcal{B}_k = \{B \subset S_{L^2} \subset H_T^1; i_{S^1}(B) \geq k\}$ to obtain

$$\nu_k(\gamma) = \inf_{\mathcal{B}_k} \sup_B J_\gamma(u). \tag{6}$$

For later considerations, we note that from the variational characterization follows immediately that $\nu_k(\gamma)$ is decreasing in γ; furthermore, one has $\nu_k(0) = \lambda_k$, the $k-th$ periodic eigenvalue of $-d^2/dt^2$, and $\lim_{\gamma \to +\infty} \nu_k(\gamma) = \frac{\lambda_k}{4}$.

2. MOUNTAIN-PASS THEOREMS

Another example of a class of sets which is invariant under the gradient flow is given by the *mountain-pass* theorem of A. Ambrosetti - P.H. Rabinowitz [AR]. It is based on the following: Let E be a Banach space, and $J : E \to R$ be a C^1-functional which satisfies the Palais-Smale condition. Let Γ be a family of continuous paths $\gamma : [0,1] \to E$ such that $\gamma(0) = u_0$, $\gamma(1) = u_1$, and assume that for some $\delta > 0$ *every* path $\gamma \in \Gamma$ satisfies $max_{u \in \gamma([0,1])} J(u) \geq max\{J(u_0), J(u_1)\} + \delta$. A sufficient condition for this to hold is the existence of a "mountain range" around the point u_0, i.e. the existence of a $\rho > 0$ such that $inf_{\partial B_\rho(u_0)} J(u) > max\{J(u_0), J(u_1)\}$. Again, a critical point is then given by

$$c = inf_{\gamma \in \Gamma} max_{u \in \gamma([0,1])} J(u) .$$

The invariance of the class Γ under the gradient flow is given by a *deformation lemma* which asserts that if c is not a critical value, then, for $\bar{\epsilon} > 0$ given, there exists an $\epsilon \in (0, \bar{\epsilon})$ and a homeomorphism $\eta : E \to E$ such that $\eta(\{J \leq c + \epsilon\}) \subset \{J \leq c - \epsilon\}$, and $\eta = id$ on $\{J \leq c - \bar{\epsilon}\} \cup \{J \geq c + \bar{\epsilon}\}$ (here $\{J \leq a\} = \{u \in E; J(u) \leq a\}$).

"Generalized" mountain-pass theorems [R] can be obtained as follows: Let $\Sigma_k = \{\sigma : D^k \to E \text{ continuous}; \sigma|_{\partial D^k} = id\}$, where D^k denotes a k-dimensional, closed ball in E. Then

$$c = inf_{\sigma \in \Sigma} max_{u \in \sigma(D^k)} J(u) \tag{7}$$

is a critical value provided that for every $\sigma \in \Sigma_k$ one has $max_{\sigma(D^k)} J(u) \geq max_{\partial D^k} J(u) + \delta$.

We now show that the critical values which were obtained via index-theories in the previous paragraph can also be found by mountain-pass theorems, using an inductive procedure. As an example we consider the nonlinear eigenvalue problem (2) above, that is the functional $J(u)$ given by (3). To begin, we have $\lambda_1(\rho) = inf_{\rho S_{L^2}} J(u)$. Denote by $\pm v_1 \in \rho S_{L^2}$ the corresponding first eigenfunctions. Then, defining $\Gamma_2 = \{\gamma_2 : [-1,1] \to \rho S_{L^2} \subset H_0^1$ continuous; $\gamma_2(\pm 1) = \pm v_1\}$, set

$$c_2(\rho) = inf_{\Gamma_2} sup_{\gamma_2} J(u) .$$

Now the mountain-pass procedure can be applied to the family Γ_2 provided that, for some $\delta > 0$, $sup_{\gamma_2} J(u) \geq \lambda_1 + \delta$, for every $\gamma_2 \in \Gamma_2$; in that case one obtains that $c_2(\rho)$ is a critical value, that is $c_2(\rho) = \lambda_2(\rho)$, the second (variational) eigenvalue of (2).

To proceed to the next step, choose for a given $\epsilon > 0$ a path $\gamma_{2,\epsilon} \in \Gamma_2$ such that $sup_{\gamma_{2,\epsilon}} J(u) \leq c_2 + \epsilon$. We now think this path defined on the semi-circle $S^1_+ = \{(x_1, x_2) \in R^2; x_1^2 + x_2^2 = 1, x_2 \geq 0\}$, and extend it to the full circle $S^1 \subset R^2$ by oddness, i.e. we set $\tilde{\gamma}_{2,\epsilon} : S^1 \to S_{L^2}, \tilde{\gamma}_{2,\epsilon} = \gamma_{2,\epsilon}$ on S^1_+ and $\tilde{\gamma}_{2,\epsilon} = -\gamma_{2,\epsilon}$ on $S^1_- = -S^1_+$. Then set $\Gamma_3 = \{\gamma_3 : D^2 \subset R^2 \to S_{L^2}$ continuous; $\gamma_3|_{\partial D^2} = \tilde{\gamma}_{2,\epsilon}\}$, where D^2 denotes the unit disk in R^2. Note that this family depends on ϵ. If now

$$c_3(\rho, \epsilon) := inf_{\Gamma_3} sup_{\gamma_3} J(u) > c_2(\rho),$$

then $c_3(\rho, \epsilon)$ is a critical value which does not depend on ϵ, i.e. $c_3(\rho, \epsilon) = \lambda_3(\rho)$. It would be of interest to show that if this condition is not satisfied, then there is a continuum of solutions at the level $\lambda_2(\rho)$ (as can be done via the genus).

To continue, choose again an element $\gamma_{3,\epsilon} \in \Gamma_3$ such that $sup_{\gamma_{3,\epsilon}} \leq c_3 + \epsilon$. Then, thinking this mapping defined on the hemi-sphere $S^2_+ = \{(x_1, x_2, x_3) \in R^3; \sum_{i=1}^{3} x_i^2 = 1, x_3 \geq 0\}$, we extend it again by oddness to the mapping $\tilde{\gamma}_{3,\epsilon} : S^2 \to \rho S_{L^2}$. Defining the family $\Gamma_4 = \{\gamma_4 : D^3 \subset R^3 \to \rho S_{L^2}$ continuous; $\gamma_4|_{\partial D^3} = \tilde{\gamma}_{3,\epsilon}\}$, one sets $c_4(\rho, \epsilon) = inf_{\Gamma_4} sup_{\gamma_4} J(u)$, and so on, i.e. in general one has

$$c_k(\rho, \epsilon) = inf_{\Gamma_k} sup_{\gamma_k} J(u) . \tag{8}$$

The same approach works also for the S^1-symmetry considered above. In fact, in the example of the periodic Fučik-spectrum, one begins with $\nu_1(\gamma) = inf_{S_{L^2}} J_\gamma(u)$, where now S_{L^2} denotes the unit L^2-sphere in H^1_T. Note that the minimizer is given by the constant function $\mathbf{1} = 1/T^{1/2}$, and that $J(-\mathbf{1}) = 0$. Defining as before the class $\Sigma_2 = \{\sigma_2 : [-1, 1] \to S_{L^2} \subset H^1_T$ continuous; $\sigma_2(\pm 1) = \pm \mathbf{1}\}$, one has

$$\nu_2(\gamma) = inf_{\Sigma_2} sup_{\sigma_2} J_\gamma(u) ,$$

since this value is strictly positive. Now choose, for an $\epsilon > 0$, a $\sigma_{2,\epsilon} \in \Sigma_2$ with $sup_{\sigma_{2,\epsilon}} J_\gamma(u) \leq \nu_2(\gamma) + \epsilon$. We think this function defined on the semi-circle $S^1_z = \{(s, tz) \in R \times C; s^2 + t^2 = 1, t \geq 0\}$, where z is a given complex number with $|z| = 1$. We extend $\sigma_{2,\epsilon}$ to the mapping $\tilde{\sigma}_{2,\epsilon} : S^2 = \{(s, y) \in R \times C; s^2 + |y|^2 = 1\} \to S_{L^2}$ by using the natural S^1-action $e^{i\theta}, \theta \in [0, 2\pi)$, on the complex numbers C; namely, we set $\tilde{\sigma}_{2,\epsilon}(s, te^{i\theta}z) = T_\theta \sigma_{2,\epsilon}(s, tz)$, where T_θ is the S^1-action by time shifts introduced above. Note that due to the

T_θ–invariance of $J_\gamma(u)$ we have $sup_{\sigma_{2,\epsilon}} J_\gamma(u) = sup_{\tilde\sigma_{2,\epsilon}} J_\gamma(u)$. The next steps are done analogously, and one has again that for $\epsilon > 0$ sufficiently small

$$\nu_k(\gamma) = inf_{\Sigma_k} sup_{\sigma_k} J_\gamma(u) .$$

3. PERTURBATION OF SYMMETRY

The advantage of the mountain-pass approach to symmetric problems is that it works also for functionals with *perturbed symmetry*. Such problems have been considered by Struwe [S], Bahri-Berestycki [BB] and Rabinowitz [R1]. As an example let us consider the problem of finding variationally the solutions of

$$-\Delta u + g(x,u) = \lambda u + f , \text{ in } \Omega , u = 0 \text{ on } \partial\Omega , \qquad (9)$$

where $f \in L^2(\Omega)$ is a given forcing term, and $g : \overline\Omega \times R \to R$ is given as in example (2), i.e. continuous and odd; furthermore, we assume that $\frac{1}{t^2} G(x,t) \to \frac{1}{2}\beta(x)$ as $|t| \to \infty$, where G denotes again a primitive of g, and $\beta \in C(\overline\Omega)$. The corresponding functional $I_f : H_0^1(\Omega) \to R$

$$I_f(u) = \frac{1}{2}\int_\Omega |\nabla u|^2 + \int_\Omega G(x,u) - \frac{\lambda}{2}\int_\Omega u^2 - \int_\Omega fu$$

is cleary not symmetric, but we can consider it as a perturbation of the eigenvalue problem (2) above.

Indeed, let $c_k(\rho,\epsilon)$ denote the values given by (8) for the symmetric functional $J(u) = \frac{1}{2}\int_\Omega |\nabla u|^2 + \int_\Omega G(x,u)$, and set

$$a_k(\epsilon) = lim_{\rho \to +\infty} \frac{1}{\rho^2} c_k(\rho,\epsilon) , b_{k+1}(\epsilon) = lim_{\rho \to \infty} \frac{1}{\rho^2} c_{k+1}(\rho,\epsilon).$$

Then we have the following

1. Theorem: Assume that for some $k \in N$ and some $\epsilon > 0$ we have $a_k(\epsilon) < \lambda < b_{k+1}(\epsilon)$, and assume that the corresponding functional I_f satisfies the Palais-Smale condition. Then equation (9) has a solution for every $f \in L^2(\Omega)$.

Proof: The proof relies on the following estimate: Define the family $\Gamma = \{\gamma : D^k \subset R^k \to \rho S_{L^2} \subset H_0^1 \text{ continuous}; \gamma|_{\partial D^k} = \rho\tilde\gamma_{k,\epsilon}\}$, where $\rho > 0$ will be suitably chosen, and $\tilde\gamma_{k,\epsilon}$ is the approximating (and extended) function for $c_k(\rho,\epsilon)$ obtained inductively for J (see above).

We now claim that for ρ sufficiently large

$$c = inf_\Gamma sup_\gamma I_f(u)$$

is a critical value. For this, we verify that $sup_\gamma I_f(u) \geq -\,const.$, for every $\gamma \in \Gamma$ and independently of $\rho \geq \rho_0 > 0$; in fact, setting $\delta = \frac{1}{4}min\{\lambda - a_k, b_{k+1} - \lambda\}$, we choose $\rho_0 > 0$ such that $\frac{1}{\rho^2}c_{k+1}(\rho, \epsilon) \geq b_{k+1}(\epsilon) - \delta$, $\frac{1}{\rho^2}c_k(\rho, \epsilon) \leq a_k(\epsilon) + \delta$, for all $\rho \geq \rho_0$, and $|\int_\Omega (G(x,u) - \frac{1}{2}\beta(x)u^2)| \leq \delta \int_\Omega u^2 + c$, for all $\int_\Omega u^2 \geq \rho_0$. Then one can estimate

$sup_{\gamma(D^k)} I_f(u)$

$\geq sup_{\gamma(D^k)} \{\frac{1}{2}\int_\Omega |\nabla u|^2 + \int_\Omega G(x,u) - \frac{\lambda}{2}\int_\Omega u^2 - \delta \int_\Omega u^2\} - c(\delta)\int_\Omega f^2$

$\geq sup_{\gamma(D^k)} \frac{1}{2}\{\int_\Omega |\nabla u|^2 + (\beta(x) - \lambda - 3\delta)u^2\} - c =: sup_{\gamma(D^k)} Q(u) - c$.

We claim that the last expression is bounded below. This is clearly the case if $0 \in \gamma(D^k)$. If $0 \notin \gamma(D^k)$, then we note that $sup_{p_\rho \circ \gamma(D^k)} Q(u) = Q(u_k) \geq \delta \rho^2 \int_\Omega \bar{u}_k^2$ (here $p_\rho : E \setminus \{0\} \to \rho S_{L^2}$, $u \mapsto \rho \frac{u}{\|u\|} =: \rho\bar{u}$, is the radial projection); this is the case by the variational characterization (8), since $p_\rho \circ \gamma(D^k) \in \Gamma_{k+1}$. Thus $sup_{\gamma(D^k)} Q(u) \geq \delta \|u_k\|^2$, and hence the claim.

On the other hand, for $\rho \to +\infty$ we have

$sup_{\gamma(\partial D^k)} I_f(u) = sup_{\rho\gamma_k, \epsilon} I_f(u)$

$\leq \frac{1}{2}(a_k + \delta - \lambda + 2\delta)\rho^2 \int_\Omega \bar{u}^2 + c(\delta)\int_\Omega f^2 \to -\infty,$

Hence we can choose $\rho_1 \geq \rho_0$ such that for $\rho \geq \rho_1$ the hypothesis for the generalized mountain-pass theorem (7) are satisfied, and c is critical.

We emphasize that the asymptotic growth assumption in theorem 1 is made on the primitive G of the nonlinearity g. As an illustration of this, and an application of theorem 1, we consider the following

Example: Let $\lambda_k, k \in N$, denote the eigenvalues of the Laplacian. Assume that $\lambda_k < \lambda < \lambda_{k+1}$, for some $k \in N$, and let $a \in R$ s.th. $|a| \leq max\{\lambda_{k+1} - \lambda, \lambda - \lambda_k\}$. Then the equation

$$-\Delta u - \lambda u + au\cos(u) = f \text{ in } \Omega, \quad u = 0 \text{ on } \partial\Omega, \qquad (10)$$

has a solution for every forcing term $f \in L^2(\Omega)$.

Proof: First note that critical points of the functional $I_f : H_0^1(\Omega) \to R$,

$$I_f(u) = \frac{1}{2}\int_\Omega |\nabla|^2 - \frac{1}{2}\int_\Omega u^2 + a\int_\Omega (\cos(u) + u\sin(u)) - \int_\Omega fu$$

yield solutions of (10). We verify the hypothesis of theorem 1:
i) We first construct suitable values $c_k(\rho)$ for the symmetric functional $J(u) = \frac{1}{2}\int_\Omega |\nabla|^2 + a\int_\Omega(cos(u) + u sin(u))$. Let $j : R^{k-1} \to E^{k-1}, j((x_1, ..., x_{k-1})) = \sum_{i=1}^{k-1} |x_i|e_i$, denote the natural embedding of R^{k-1} onto the space E^{k-1} spanned by the first $k-1$ eigenfunctions e_i of the Laplacian (we assume that $\|e_i\|_{L^2} = 1$), and define the families

$$\Gamma_k(\rho) = \{\gamma : D^{k-1} \subset R^{k-1} \to \rho S_{L^2} \subset H_0^1(\Omega); \gamma \text{ continuous and}$$
$$\text{such that } \gamma|_{\partial D^{k-1}} = \rho \cdot j\}.$$

We can estimate

$$c_k(\rho) := inf_{\Gamma_k(\rho)} sup_{\gamma(D^{k-1})} J(u) \leq sup_{j(D^k)} J(u) \leq \lambda_k \rho^2 + a \cdot c \cdot \rho.$$

On the other hand, note that every $\gamma \in \Gamma_k(\rho)$ intersects $S_{L^2} \cap (E^{k-1})^\perp$. Otherwise we could consider γ defined on $D_+^k = \{(x_1, ..., x_k) \in R^k; \sum_{j=1}^k = 1, x_k \geq 1\}$ and extend it by oddness to a continuous and odd map $\tilde{\gamma} : D^k \to S_{L^2} \cap E^{k-1}$; but this contradicts the *Borsuk-Ulam* theorem. Hence, for every $\gamma \in \Gamma_k(\rho)$ we can find a $x_k \in (E^{k-1})^\perp$ such that $c_k(\rho) \geq \frac{1}{2}\rho^2 \int_\Omega |\nabla x_k|^2 - a \cdot c \cdot \rho$, and hence

$$c_k(\rho) \geq \frac{1}{2}\rho^2 \lambda_k - a \cdot c \cdot \rho.$$

Thus we find $\frac{1}{\rho^2} c_k(\rho) \to \lambda_k$, i.e. the first hypothesis of theorem 1 is satisfied.

ii) It remains to verify that the functional I_f satisfies the Palais-Smale condition, i.e. we have to verify that any sequence $\{u_n\} \subset H_0^1$ which satisfies $|I_f(u_n)| \leq c$ and $I_f'(u_n) \to 0$ in H^{-1} contains a convergent subsequence in $H_0^1(\Omega)$. Let $\{u_n\}$ be such a sequence. Taking the difference $I_f'(u_n)[u_n] - 2I_f(u_n)$ we conclude that

$$\left|\int_\Omega u_n^2 cos(u_n)\right| \leq c\|u_n\|_{L^2}.$$

We first show that $\|u_n\|_{L^2}$ is bounded; assume that this is not so, and set $\bar{u}_n := u_n/\|u_n\|_{L^2}$. From $|I_f(u_n)| \leq c$ we obtain that $\int_\Omega |\nabla \bar{u}_n|^2 \leq c$, and hence there exists a convergent subsequence $\bar{u}_n \to \bar{u}$ in $L^2(\Omega)$. With this one now derives the following (weak) limit equation:

$$-\Delta \bar{u} - \lambda \bar{u} + \bar{u} \cdot y = 0.$$

Since by the above estimate $\int_\Omega \bar{u}^2 y = 0$ we conclude that y changes sign, and since $\lambda_k \leq \lambda - y \leq \lambda_{k+1}$, with $\lambda - y \not\equiv \lambda_j, j = 1, 2$, the equation

$-\Delta \bar{u} = (\lambda - y)\bar{u}$ implies $\bar{u} = 0$. This contradicts $\|\bar{u}\|_{L^2} = 1$, and hence u_n is bounded in L^2. From this the convergence in $H_0^1(\Omega)$ follows easily.

While in theorem 1 the growth ot G may be at most quadratic, we show in the following theorem that such a restriction is not required in the case of periodic solutions of Sturm-Liouville equations; indeed, the typical example for the nonlinearity in the following theorem is $g(u) = exp(u)$. This problem was considered in a recent joint work with D.G. deFigueiredo [FR]:

$$-u'' - \lambda u = g(t, u) + f(t) , \quad u \; T-periodic , \qquad (11)$$

where $g(t, u)$ is continuous, T−periodic in t, superlinear at $+\infty$ and sublinear at $-\infty$. Under some additional conditions on g (which are used to prove the Palais-Smale condition), it was proved in [FR] that:

2. Theorem: If $\lambda > 0$ then (11) has a solution for *every* $f \in L^2(0,T)$.

Note that if $\lambda \leq 0$ then there exist forces $f \in L^2(0,T)$ for which (10) has no solution; a necessary condition is obtained by integrating equation (10) over a time T interval.

The *proof* of theorem 2 relies again on a *comparison* of critical levels. Indeed, the functional

$$J(u) = \frac{1}{2}\int_0^T |u'|^2 - \frac{\lambda}{2}\int_0^T u^2 - \int_0^T G(t,u) - \int_0^T fu$$

corresponding to (10) can be considered (somewhat surprisingly) as a perturbation of the functional $J_\gamma(u)$ given by (5), with γ sufficiently large. In fact, we proceed again inductively. Assume first that $0 < \lambda < \frac{\lambda_k}{4}$. Then let $\mathcal{L}_1 = \{h : S_+^1 \to E \text{ continuous}; h((\pm 1, 0)) = \pm r\mathbf{1}\}$. One obtains (using that $J(\pm r\mathbf{1}) \to -\infty$ as $r \to +\infty$) that for $r > 0$ sufficiently large (and $\delta = \frac{1}{2}(\frac{\lambda_1}{4} - \lambda)$)

$$sup_{h(S_+^1)} J(u)$$
$$\geq sup_{h(S_+^1)} \frac{1}{2} J_\gamma(u) - \frac{1}{2}(\lambda + \delta)\int_0^T u^2 - c(\delta)\int_0^T f^2$$
$$> J(\pm r\mathbf{1}) + 1,$$

for every $h \in \mathcal{L}_1$.

For the next level $\lambda \in (\frac{\lambda_1}{4}, \frac{\lambda_2}{4})$ one first chooses γ so large that $\nu_1(\gamma) < \lambda$ (this is possible since $\nu_1(\gamma) \to \frac{\lambda_1}{4}$ as $\gamma \to +\infty$) and selects a map $\bar{\gamma}_{1,\epsilon} : D^2 \to S_{L^2} \subset H_T^1$ (see above) such that $sup_{\bar{\gamma}_{1,\epsilon}} J_\gamma(u) \leq \nu_1(\gamma) + \epsilon$. One then estimates that $\sup_{r\bar{\gamma}_{1,\epsilon}} J(u) \to -\infty$ as $r \to \infty$. Setting $\mathcal{L}_2 = \{h : S_+^2 \to H_T^1$ continuous; $h|_{S^1} = r\bar{\gamma}_{1,\epsilon}\}$, one proves again that for r sufficiently large and $\delta \leq \frac{1}{2}(\frac{\delta_2}{4} - \lambda)$ one has

$$\sup_{h(S_+^2)} J(u) > \sup_{h(S^1)} J(u) + 1,$$

for every $h \in \mathcal{L}_2$. The minimax principle then yields again that

$$c_2 = \inf_{h \in \mathcal{L}_2} \sup_{u \in h} J(u)$$

is a critical value.

The cases $\frac{\lambda_k}{4} < \lambda < \frac{\lambda_{k+1}}{4}$ are treated, inductively, in the same manner.

The above procedure does not work for $\lambda = \frac{\lambda_k}{4}, k \in N$; however, choosing $\lambda_{(n)} \in (\frac{\lambda_k}{4}, \frac{\lambda_{k+1}}{4})$ with $\lambda_{(n)} \to \frac{\lambda_k}{4}$ as $n \to \infty$ and corresponding solutions u_n, one sees that these solutions converge to a solution $\bar{u}$ of (10) with $\lambda = \frac{\lambda_k}{4}$.

It would be of interest to extend the above result to problems with other boundary conditions and to equations in higher dimensions.

REFERENCES

[AR] Ambrosetti, A., Rabinowitz, P.H., *Dual variational methods in critical point theory and applications*, J. Funct. Anal. **14** (1973) 349-381.

[B] Benci, V., *A geometrical index for the group S^1 and some applications to the study of periodic solutions of ordinary differential equations*, Comm. Pure Appl. Math. **34** (1981) 393-432

[BB] Bahri, A., Berestycki, H., *A perturbation method in critical point theory and applications*, Trans. Amer. Math. Soc. **267** (1981) 1-32

[BLMR] Berestycki, H., Lasry, J.M., Mancini, G., Ruf, B., *Existence of multiple periodic orbits on star-shaped Hamiltonian surfaces*, Comm. Pure Appl. Math. **38** (1985) 253-289

[C] Coffman, C.V., *A minimum-maximum principle for a class of nonlinear integral equations*, J. Analyse Math. **22** (1969) 391-419

[FR] Figueiredo, D.G., Ruf, B., *On the periodic Fučik spectrum and a superlinear Sturm-Liouville equation*, Proc. Royal Soc. Edinb. **123 A** (1992)

[K] Krasnoselskii, M.A., *Topological methods in the theory of nonlinear integral equations*, Macmillan, New York (1964)

[R] Rabinowitz, P.H., *Some critical point theorems and applications to semilinear elliptic partial differential equations*, Ann. Sc. Norm. Sup. Pisa, Ser. 4, **5** (1978) 215-223

[R1] Rabinowitz, P.H., *Multiple critical points of perturbed symmetric functionals*, Trans. Amer. Math. Soc. **272** (1982) 753-770

[S] Struwe, M., *Infinitely many critical points for functionals which are not even and applications fo superlinear boundary value problems*, Manusc. math. **32** (1980) 335-364

[S1] Struwe, M., *Variational methods*, Springer Verlag, Berlin Heidelberg, 1990

A TWO POINTS BOUNDARY VALUE PROBLEM ON NON COMPLETE RIEMANNIAN MANIFOLDS

A. SALVATORE

Dipartimento di Matematica,
Universitá degli Studi di Bari, via E. Orabona, 4
70125 Bari - Italy

§ 0. INTRODUCTION

Let $(\mathfrak{M}, <\cdot,\cdot>_R)$ be a Riemannian manifold, $x_0, x_1 \in \mathfrak{M}$ and $V: \mathfrak{M} \to \mathbb{R}$ a C^1 potential function. In this paper we fix $T > 0$ and we look for curves $x: [0,T] \to \mathfrak{M}$ (T>0) satisfying the equation

$$D_t \dot{x}(t) = -\nabla_R V(x(t)) \qquad (0.1)$$

with boundary conditions

$$x(0) = x_1, \qquad x(T) = x_2 \qquad (0.2)$$

where $D_t \dot{x}(t)$ denotes the covariant derivative of $\dot{x}(t)$ along the direction of $\dot{x}(t)$ and $\nabla_R V(x(t))$ is the Riemanniann gradient with respect to x of V in $(x(t))$.

The existence of infinitely many solutions of (0.1)-(0.2) has been proved in [12] if $\mathfrak{M}$ is complete and V is locally Lipschitz continuous.

If V=0, problem (0.1)-(0.2) reduces to the classical geodesics problem. Existence and multiplicity results of geodesics or closed geodesics have been stated when the manifold is compact or at least complete (see [16], [10], [5], [17]).

In this paper we deal with a non complete Riemannian manifold. Let d denote the canonical distance on $\mathfrak{M}$ induced by the

Riemannian structure and let $T_x(\mathfrak{M})$ be the tangent space to $\mathfrak{M}$ at x.

We shall introduce the following definition (see [4], [11]):

DEFINITION 0.1 Let $\mathfrak{M}$ be an open connected subset of a Riemannian manifold $\overline{\mathfrak{M}}$ and $\partial \mathfrak{M}$ its topological boundary. $\mathfrak{M}$ is said to be a Riemannian manifold with convex boundary if there exists a positive map $\Phi \in C^2(\mathfrak{M}, \mathbb{R}_+ \setminus \{0\})$ having the following properties:

i) $\Phi(x) \to 0$ as $x \to \bar{x} \in \partial \mathfrak{M}$.

ii) For any $\eta > 0$ the set $\{x \in \mathfrak{M} | \Phi(x) \geq \eta\}$ is complete (with respect to the Riemannian structure of $\mathfrak{M}$).

iii) There exist some positive constants α, β, γ, δ such that for any $x \in \mathfrak{M}$ with $\Phi(x) < \delta$ it results:
$$\alpha \leq \langle \nabla_R \Phi(x), \nabla_R \Phi(x) \rangle_R \leq \beta$$
and
$$H_R^\Phi(x)[v,v] \leq \gamma \langle v, v \rangle_R \Phi(x) \qquad \text{for any } v \in T_x(\mathfrak{M})$$
where $H_R^\Phi(x)[v,v]$ denotes the Riemannian Hessian of Φ at x in the direction v.

Let us point out that by i) and iii) it follows that
$$\limsup_{x \to \bar{x} \in \partial \mathfrak{M}} H_R^\Phi(x)[v,v] \leq 0 \qquad \text{for any } v \in T_x(\mathfrak{M}) \text{ with } \langle v, v \rangle_R \leq 1.$$

The convexity assumption in definition 0.1 permits to prove the existence of solutions of (0.1) with boundary conditions even if the manifold $\mathfrak{M}$ is not complete. We recall that if $V = 0$, the existence of geodesics on a Riemannian manifold with boundary has been proved by Gordon under stronger convexity assumptions (see [8], [9]).

In this paper we shall study problem (0.1)-(0.2) by variational methods, indeed it is known that the solutions of equation (0.1) with boundary conditions are the critical points of the functional
$$f(x) = \int_0^T \langle \dot{x}, \dot{x} \rangle_R \, dt - \int_0^T V(x) \, dt$$
defined on a suitable Riemannian manifold (see section 1).

We shall prove the following theorem:

THEOREM 0.2 Let $(\mathfrak{M},<\cdot,\cdot>_R)$ be a finite dimensional C^3 Riemannian manifold with convex boundary. Let $V:\mathfrak{M} \to \mathbb{R}$ be a C^1 function satisfying the following assumptions:

(V_1) There exists $c_0 \in \mathbb{R}$ such that
$$V(x) < c_0 \quad \text{for any } x \in \mathfrak{M},$$

(V_2) there exists $c^* \in \mathbb{R}$, $c^* < \alpha$, such that

i) $V(x) \geq - c^*/\Phi^4(x)$ for any $x \in \mathfrak{M}$ with $\Phi(x) < \delta$

ii) $<\nabla_R \Phi(x), \nabla_R V(x)>_R \geq 0$ for any $x \in \mathfrak{M}$ with $\Phi(x) < \delta$

(where δ is the constant introduced in (iii) of definition 0.1).

Then for any $x_1, x_2 \in \mathfrak{M}$ and for any $T > 0$, there exists at least a solution of problem (0.1)-(0.2).

Moreover, if $\mathfrak{M}$ is not contractible in itself, then problem (0.1)-(0.2) admits infinitely many solutions x_n such that $f(x_n) \to +\infty$ as $n \to +\infty$.

REMARK 0.3 Let us point out that if $\mathfrak{M}$ is a Riemannian manifold with convex boundary, the existence of infinitely many T-periodic solutions of the equation (0.1) has been stated in [14].

§ 1. PRELIMINARY AND NOTATIONS

Let $(\mathfrak{M},<\cdot,\cdot>_R)$ be a finite dimensional C^3 Riemannian manifold with convex boundary and let d denote the canonical distance induced by the Riemannian structure.

Set

$$W^{1,2}([0,T]),\mathfrak{M})=\{x:[0,T]\to\mathfrak{M}, \ x \text{ absolutely continuous}, \int_0^T <\dot{x},\dot{x}>_R dt <+\infty\}$$

$$\Omega^1 = \Omega^1(\mathfrak{M},x_1,x_2) = \{x \in W^{1,2}([0,T],\mathfrak{M}) \mid x(0) = x_1, \ x(T) = x_2\}$$

It is known that Ω^1 is a Riemannian manifold (see [15],[10]) and its tangent space at $x \in \Omega^1$ is given by

$$T_x\Omega^1 = \{\xi \in W^{1,2}([0,T],T\mathfrak{M}) \mid \xi(0)=\xi(T)=0 \text{ and } \xi(t) \in T_{x(t)}\mathfrak{M} \ \forall t \in [0,T]\}$$

where $T\mathfrak{M}$ is the tangent bundle of $\mathfrak{M}$ and $W^{1,2}([0,T],T\mathfrak{M})$ is the set of the absolutely continuous curves $\xi:[0,T] \to T\mathfrak{M}$ such that

$$\langle\xi,\xi\rangle_1 = \int_0^T \langle D_t\xi(t), D_t\xi(t)\rangle_R \, dt < +\infty .$$

Let $C([0,T],\mathfrak{M}))$ denote the space of the continuous curves in $\mathfrak{M}$ endowed with the metric

$$d_\infty(x,x') = \sup_{t\in[0,T]} d(x(t),x'(t)) .$$

It is easy to see that the solutions of problem (0.1) with boundary conditions (0.2) are the critical points of the following C^1- action functional

$$f(x) = \frac{1}{2}\int_0^T \langle \dot{x},\dot{x}\rangle_R \, dt - \int_0^T V(x) \, dt \quad (1.1)$$

defined on Ω^1.

Unfortunately, since $\mathfrak{M}$ is not complete, also Ω^1 is not complete, and therefore f does not satisfy the classical Palais-Smale condition, which is very important in order to apply the classical results of the critical point theory.

Then, we shall penalize the functional f in a suitable way and we shall prove that the penalized functional has critical points which are also critical points of f.

Let us recall now the Palais-Smale condition for a functional:

DEFINITION 1.1 Let $\mathfrak{N}$ be a Riemannian manifold and $f:\mathfrak{N} \to \mathbb{R}$ a C^1 functional; f is said to satisfy the Palais-Smale condition iff

(P.S) $\begin{cases} \text{Any sequence } \{x_n\} \subset \mathfrak{N} \text{ such that} \\ \{f(x_n)\} \text{ is bounded} \hfill (1.2) \\ \{f'(x_n)\} \to 0 \text{ as } n \to +\infty , \hfill (1.3) \\ \text{admits a convergent subsequence in } \mathfrak{N}. \end{cases}$

A sequence satisfying (1.2)-(1.3) is named a (P.S) sequence.

§ 2. STATEMENT OF THE RESULT

Now, we look for the solutions of (0.1)-(0.2), i.e. the critical points of the energy functional f on Ω^1.

As $\mathfrak{M}$ is not complete, the functional f does not satisfy (P.S) because any (PS) sequence is bounded in Ω^1 but it can be converge to an element $x \in \partial\Omega^1$.

This difficulty arises also in the study of geodesics on Lorentian manifold with convex boundary, then, adapting the arguments used in [4], we shall consider for any $\varepsilon > 0$ the following penalized functional defined on Ω^1:

$$f_\varepsilon(x) = \frac{1}{2}\int_0^T <\dot{x},\dot{x}>_R dt - \int_0^T V(x)dt + \int_0^T U_\varepsilon(x)dt \qquad (2.1)$$

where

$$U_\varepsilon(x) = \Psi_\varepsilon\left(\frac{1}{\Phi^2(x)}\right) \qquad (2.2)$$

and $\Psi_\varepsilon : \mathbb{R} \to \mathbb{R}$ is a C^2 function satisfying

$$\begin{cases} \Psi_\varepsilon(\tau) = 0 & \text{for } \tau \leq 1/\varepsilon \\ \Psi'_\varepsilon(\tau) > 0 & \text{for } \tau \geq 1/\varepsilon \\ \Psi'_\varepsilon(\tau) = 1 & \text{for } \tau \geq 1 + 1/\varepsilon \end{cases}$$

and $\Psi_\varepsilon(\tau) \leq \Psi_{\varepsilon'}(\tau)$ for any τ and $\varepsilon \leq \varepsilon'$.

Let us recall that the following lemma holds:

<u>Lemma 2.1</u> Let $\mathfrak{M}$ be a Riemannian manifold with convex boundary. Then for any sequence $\{x_n\} \in \Omega^1$ such that

$$\int_0^T <\dot{x}_n(t),\dot{x}_n(t)>_R dt \qquad \text{is bounded}$$

and there exists $t_n \in [0,T]$ such that

$$\lim_{n \to +\infty} \Phi(x_n(t_n)) = 0 ,$$

it results

$$\lim_{n \to +\infty} \int_0^T \frac{1}{\Phi^2(x_n(t))} dt = +\infty$$

<u>Proof</u> See [4, lemma 2.3].

In the sequel we shall denote by a_i some positive constants. The following lemma justifies the introduction of the penalized functional f.

Lemma 2.2 There exists $\varepsilon_0 > 0$ such that for any $\varepsilon \in]0, \varepsilon_0]$, any critical point $x_\varepsilon \in \Omega^1$ of the functional f_ε satisfying

$$f_\varepsilon(x_\varepsilon) \le M \tag{2.3}$$

(where M is a constant indipendent of ε) is a critical point of f.

Proof For any $\varepsilon > 0$, let x_ε be a critical point of f_ε satisfying (2.3). In order to prove lemma 2.2 it is sufficient to prove that there exists a positive constant ν such that for any $\varepsilon \in [0,1]$ and for any $t \in [0,T]$ it results

$$\Phi(x_\varepsilon(t)) \ge \nu \tag{2.4}$$

In fact, if (2.4) holds, by the definition of the penalization term it follows that, for ε small enough,

$$U_\varepsilon(x_\varepsilon(t)) = 0 \qquad \text{for any } t \in [0,T]$$

and therefore x_ε is a critical point of f.

Assume by contradiction that (2.4) does not hold, then there exist $\{\varepsilon_n\} \to 0$ and $\{x_n\} \subset \Omega^1$ such that x_n is a critical point of $f_n = f_{\varepsilon_n}$ satisfying (2.3) and

$$\Phi(x_n(t_n)) \to 0 \qquad \text{as } n \to +\infty \tag{2.5}$$

where t_n is a minimum point of the map $v_n(t) = \Phi(x_n(t))$.

As x_n is a critical point of f_n, it results that

$$D_t \dot{x}_n = -\nabla_R V(x_n) - 2\Psi'_{\varepsilon_n}\left(\frac{1}{\Phi^2(x_n)}\right) \frac{\nabla_R \Phi(x_n)}{\Phi^3(x_n)} \tag{2.6}$$

By (2.3) and (V_1) it follows that $\{\int_0^T \langle \dot{x}_n, \dot{x}_n \rangle_R \, dt\}$ is bounded, then there exist a subsequence of $\{x_n\}$, still denoted $\{x_n\}$, and $x \in W^{1,2}([0,T], \mathbb{R}^N)$ such that $x_n \to x$ weakly in $W^{1,2}$ and uniformly in $[0,T]$.

NON COMPLETE RIEMANNIAN MANIFOLDS 155

By (2.5) $x \in \partial \Lambda^1$, i.e. there exists $t_0 \in [0,T]$ such that $\Phi(x(t_0))=0$ where, passing eventually to a subsequence, $t_0 = \lim t_n$.

Then, there exists $a_1 > 0$ such that for n large,

$$v_n(t) = \Phi(x_n(t)) < \delta \quad \text{for any } t \in [t_0-a_1, t_0+a_1] \quad (2.7)$$

where $\delta > 0$ is the constant introduced in (iii) of definition 0.1.

Now, as t_n is a minimum point for v_n, for any $n \in \mathbb{N}$ it results

$$\dot{v}_n(t_n) = 0, \quad \ddot{v}_n(t_n) \geq 0$$

then

$$H_R^\Phi(x_n(t_n))[\dot{x}_n(t_n), \dot{x}_n(t_n)] + \langle \nabla_R \Phi(x_n(t_n)), D_t \dot{x}_n(t_n) \rangle_R \geq 0 \quad (2.8)$$

By (2.6) and (2.8) it follows that

$$0 \leq H_R^\Phi(x_n(t_n))[\dot{x}_n(t_n), \dot{x}_n(t_n)] - \langle \nabla_R \Phi(x_n(t_n)), \nabla_R V(x_n(t_n)) \rangle_R +$$

$$-2\Psi'_{\varepsilon_n}\left(\frac{1}{\Phi^2(x_n(t_n))}\right) \langle \nabla_R \Phi(x_n(t_n)), \nabla_R \Phi(x_n(t_n)) \rangle_R \frac{1}{\Phi^3(x_n(t_n))} \quad (2.9)$$

As $\mathfrak{M}$ has convex boundary and (V_2)-ii) holds, by (2.7) and (2.9) we obtain that, for n large and for any $t \in [t_0-a_1, t_0+a_1]$,

$$0 \leq \ddot{v}_n(t) \leq \gamma \langle \dot{x}_n(t), \dot{x}_n(t) \rangle_R v_n(t) - 2\Psi'_{\varepsilon_n}\left(\frac{1}{v_n^2(t)}\right)\frac{\alpha}{v_n^3(t)} \quad (2.10)$$

Furthermore, since x_n is a critical point of f_n, there exists a constant c_n such that

$$\frac{1}{2}\langle \dot{x}_n(t), \dot{x}_n(t) \rangle_R + V(x_n(t)) - \Psi_{\varepsilon_n}\left(\frac{1}{\Phi^2(x_n(t))}\right) = c_n \quad \text{for any } t \in [0,T]$$

and therefore for any $n \in \mathbb{N}$ and for any $t \in [0,T]$

$$\langle \dot{x}_n(t), \dot{x}_n(t) \rangle_R = 2c_n - 2V(x_n(t)) + 2\Psi_{\varepsilon_n}\left(\frac{1}{\Phi^2(x_n(t))}\right) =$$

$$= \frac{2}{T}f_n(x_n) + \frac{2}{T}\int_0^T V(x_n(\tau))d\tau - 2V(x_n(t)) +$$

$$- \frac{2}{T}\int_0^T \Psi_{\varepsilon_n}\left(\frac{1}{\Phi^2(x_n(\tau))}\right)d\tau + 2\Psi_{\varepsilon_n}\left(\frac{1}{\Phi^2(x_n(t))}\right) \leq$$

$$\leq a_2 + 2\Psi_\varepsilon\left(\frac{1}{v_n^2(t)}\right) - 2V(x_n(t)).$$

Using the last inequality in (2.10), by the definition of Ψ_ε and by (V_2)-i) we deduce that for n large

$$\ddot{v}_n(t) \leq \gamma\left(a_2 + \frac{2}{v_n^2(t)} + \frac{2 c^*}{v_n^4(t)}\right) v_n(t) - \frac{2\alpha}{v_n^3(t)} \quad \forall\, t \in [t_0-a_1, t_0+a_1]$$

It follows that, for n large,

$$0 \leq \ddot{v}_n(t) \leq a_3 v_n(t) \quad \text{for any } t \in [t_0-a_1, t_0+a_1]$$

Finally, since $\dot{v}_n(t_n)=0$, by Gronwall lemma we conclude that $v_n(t) = \Phi(x_n(t))$ converges uniformly to 0 in $[t_0-a_1, t_0+a_1]$.

Then, for large n,

$$\Psi_\varepsilon\left(\frac{1}{\Phi^2(x_n)}\right) = \frac{1}{\Phi^2(x_n)} \quad \text{for any } t \in [t_0-a_1, t_0+a_1]$$

and therefore

$$\lim_n \int_0^T \frac{1}{\Phi^2(x_n)}\, dt = +\infty$$

which contradics assumption (2.3).

In order to apply the classical arguments of the critical points theory, we shall prove the following lemma:

LEMMA 2.3 Let $\varepsilon > 0$. Under the assumptions of theorem 0.2, for any $a \in \mathbb{R}$ the sublevels

$$f_\varepsilon^a = \{x \in \Omega^1 | f_\varepsilon(x) \leq a\}$$

are complete metric spaces.

Moreover the functional f satisfies the Palais-Smale condition on Ω^1.

Proof. As the subsets

$$\left\{\int_0^T \langle \dot{x}, \dot{x}\rangle_R\, dt\, \Big|\, x \in f_\varepsilon^a\right\} \text{ and } \left\{\int_0^T \Psi_\varepsilon\left(\frac{1}{\Phi^2(x)}\right) dt\, \Big|\, x \in f_\varepsilon^a\right\}$$

are bounded, by lemma 2.1 there exists $\nu > 0$ such that

$$f_\varepsilon^a \subset \Omega^1(A_\nu) = \{x \in \Omega^1 | \Phi(x(t)) \geq \nu \text{ for any } t \in [0,T]\}$$

Thus, $\mathfrak{M}$ having convex boundary, the sublevels of f_ε are complete. In order to prove the Palais-Smale condition, let $\{x_n\}$ be a sequence in Ω^1 such that

$$f_\varepsilon(x_n) \text{ is bounded} \qquad (2.11)$$

and

$$f'_\varepsilon(x_n) \to 0 \qquad (2.12)$$

By (2.11) it follows that the sequences

$$\left\{\int_0^T \langle \dot{x}_n, \dot{x}_n \rangle_R \, dt\right\} \text{ and } \left\{\int_0^T \Psi_\varepsilon\left(\frac{1}{\Phi^2(x_n)}\right) dt\right\}$$

are bounded and therefore, passing eventually to a subsequence,

$$x_n \to x \quad \text{weakly in } \Omega^1.$$

Moreover by (2.11) there exists $\nu > 0$ such that $\{x_n\} \subset \Omega^1(A_\nu)$. As $(A_\nu, \langle \cdot, \cdot \rangle_R)$ is a C^3 complete finite dimensional Riemannian manifold, by a Nash embedding theorem (see [13]) it is isometrically embedded into $\mathbb{R}^N$, with N large enough, equipped with the Euclidean metric. Then by (2.12), using lemma (2.1) of [2] and arguing in a standard way, it is possible to show that $x_n \to x$ in W^1 (see also [3, Theorem 1.1]).

Clearly, $x \in \Omega^1(A_\nu)$ and then $x \in \Omega^1$.

Now, let us recall that, if A is a closed subset of Ω^1, $\text{cat}_{\Omega^1}(A)$ denotes the Ljusternik-Schnirelman category of A in Ω^1, (see [15]), that is the minimal number of closed contractible subsets in Ω^1 covering A.

In the following lemma we will prove that the sublevels of f have finite category, even if f does not verify (P.S).

<u>Lemma</u> 2.4 For any $c \in \mathbb{R}$

$$\text{cat}_{\Omega^1}(f_c) < +\infty \qquad (2.13)$$

Proof. Let $\eta > 0$, η so small that x_1, $x_2 \in A_\eta$, where $A_\eta = \{x \in \mathfrak{M} | \phi(x) \geq \eta\}$. As $\mathfrak{M}$ has convex boundary, using the curves of maximal slope for the functional Φ, we can retract $\mathfrak{M}$ on the set A_η, that is we can construct a diffeomorphism $\psi: \mathfrak{M} \to A_\eta$.

Moreover, as ψ is Lipschitz continuous and V is bounded, it is easy to see that

$$\forall c \in \mathbb{R} \; \exists \; \sigma = \sigma(c) > 0 \text{ s.t. } \forall \; x \in f^c: \psi(x) \in f^{c+\sigma, \mu} \quad (2.14)$$

where

$$f^{c+\sigma, \mu} = \{x \in f^{c+\sigma} | x(t) \in A_\mu \quad \forall \; t \in [0,T]\} \quad (2.15)$$

Now, consider the penalized functional

$$f_*(x) = f(x) + \int_0^T \frac{\theta(x)}{\Phi^2(x)} dt \quad x \in \Omega^1$$

where $\theta(x)$ is a C^2 positive scalar field on $\mathfrak{M}$ such that

$$\theta(x) = \begin{cases} 0 & \text{for } x \in \mathfrak{M}, \; \Phi(x) \geq \mu/2 \\ 1 & \text{for } x \in \mathfrak{M}, \; \Phi(x) < \mu/3 \end{cases} \quad \theta(x) \in [0,1] \text{ for any } x \in \mathfrak{M}$$

Obviously, for any $x \in \Omega^1(A_\mu)$ we have

$$f_*(x) = f(x)$$

then

$$f^{c+\sigma, \mu} \subset f_*^{c+\sigma} = \{x \in \Omega^1 | f_*(x) \leq c+\sigma\} \quad (2.16)$$

By (2.14) and (2.16) it follows that

$$\psi(f^c) \subset f_*^{c+\sigma} \quad (2.17)$$

Now, as f_* is bounded from below and satisfies Palais-Smale condition on Ω^1 (see lemma 2.3), it can be proved that (see [3], theorem 1.1)

$$\text{cat}_{\Omega^1}(f_*^{c+\sigma}) < +\infty \quad (2.18)$$

By (2.17) and (2.18) and by well known properties of the Ljusternik-Schnirelman category, we conclude that (2.13) holds.

PROOF OF THEOREM 0.2

As for any $\varepsilon > 0$, the functional f_ε is bounded from below, satisfies (P.S) condition and its sublevels are complete, then it attains its minimum at a point $x_\varepsilon \in \Omega^1$. Furthermore,

$$f_\varepsilon(x_\varepsilon) \le f_1(x_1) \qquad \text{for any } \varepsilon \in \,]0,1]$$

Then, if ε is small enough, x_ε is a critical point of f in Ω^1. Moreover, if $\mathfrak{M}$ is not contractible in itself, there exists a sequence $\{K_n\}_n$ of compact sets of Ω^1 such that

$$\lim_{n\to+\infty} \operatorname{cat}_{\Omega^1}(K_n) = +\infty. \tag{2.19}$$

(see [6], [7]).

For any $\alpha \in \mathbb{R}$, we set

$$f_\alpha = \{x \in \Omega^1 | f(x) \ge \alpha\}, \quad f_{\varepsilon,\alpha} = \{x \in \Omega^1 | f_\varepsilon(x) \ge \alpha\}.$$

By lemma 2.4 and by (2.19) there exists $k=k(\alpha) \in \mathbb{N}$ such that

$$B \cap f_\alpha \ne \emptyset \qquad \text{for any } B \subseteq \Omega^1 \text{ with } \operatorname{cat}_{\Omega^1}(B) \ge k.$$

Then, since $f_\alpha \subseteq f_{\varepsilon,\alpha}$, it results

$$B \cap f_{\varepsilon,\alpha} \ne \emptyset \qquad \text{for any } \varepsilon > 0, \text{ for any } B \subseteq \Omega^1, \operatorname{cat}_{\Omega^1}(B) \ge k$$

and therefore

$$c_{k,\alpha} = \inf \{\sup f_\varepsilon(A) : \operatorname{cat}_{\Omega^1}(A) \ge k\} \ge \alpha. \tag{2.20}$$

Lemma 2.3 and critical points theory standard arguments imply that for any $\varepsilon \in \,]0,1]$ $c_{k,\varepsilon}$ are critical values of f_ε, that is there exists x_ε critical point of f_ε such that

$$f_\varepsilon(x_\varepsilon) = c_{k,\varepsilon} \ge \alpha. \tag{2.21}$$

Now, let $K \subseteq \Omega^1$, K compact with $\operatorname{cat}_{\Omega^1}(K) \ge k$ and $\varepsilon \in \,]0,1]$. Then

$$f_\varepsilon(x_\varepsilon) = c_{k,\varepsilon} \le \sup f_\varepsilon(K) \le \sup f_1(K) = c_1. \tag{2.22}$$

By (2.22) and lemma 2.2 we can conclude that, for ε small enough,

$$U_\varepsilon(x_\varepsilon(t)) = 0 \quad \text{for any } t$$

and therefore x_ε is a critical point of f such that $f(x_\varepsilon) \ge \alpha$.

REFERENCES

[1] V. BENCI, Periodic solutions of Lagrangian systems on a compact manifold, *J.Diff.Eq.*, 63, 135-161, (1986).

[2] V. BENCI, D. FORTUNATO, On the existence of infinitely many geodesics on space-time manifolds, to appear on *Adv. Math.*
[3] V. BENCI, D. FORTUNATO, F. GIANNONI, On the existence of multiple geodesics in static space-times, *Ann. Inst. H. Poincare', Analyse non Lineaire*, 8, 24-46, (1990).
[4] V. BENCI, D. FORTUNATO, F.GIANNONI, On the existence of geodesics in static Lorentz manifolds with singular boundary, to appear on *Ann. Scuola Norm. Sup. Pisa*.
[5] V. BENCI, F. GIANNONI, Closed geodesics on non compact Riemannian manifolds, *Compt. Rend. Acad. Sc. Paris*, 312, 857-861, (1991).
[6] A CANINO, On p-convex sets and geodesics, *J. Diff. Eq.*, 75, 118-157, (1988).
[7] E. FADELL, A. HUSSEINI. Category of loop spaces of open subsets in Euclidean space, *Nonlinear Analysis, T.M.A.*, 17, 1153-1161, (1991).
[8] W. B. GORDON, An analytical criterion for the completeness of Riemannian manifolds, *Proc. Am. Math. Soc.*, 37, 221-225, (1973).
[9] W. B. GORDON, The existence of geodesics joining two given points, *J. Diff. Geom.*, 9, 443-450, (1974).
[10] W. KLINGENBERG, Riemannian Geometry, Walter de Gruyter, Berlin / New York, 1982.
[11] A. MASIELLO, Metodi variazionali in geometria Lorenziana, Tesi di Dottorato, Pisa, (1992).
[12] E. MIRENGHI, A. SALVATORE, A non-smooth two points boundary value problem on Riemannian manifolds, to appear on *Ann. Mat. Pura e Appl.*
[13] J. NASH, The imbedding problem for Riemannian manifolds, *Ann. Math.*, 63, 20-63, (1956).
[14] A. SALVATORE, On the existence of infinitely many periodic solutions on non complete Riemannian manifolds, preprint Dip. Mat. Univ. Bari, 11, (1992).
[15] J. T. SCHWARTZ, Nonlinear Functional Analysis, Gordon and Breach, New York, 1969.
[16] J. P. SERRE, Homologie singulière des espaces fibrés, *Ann. of Math.*, 54, 425-505, (1951).
[17] THORBREGSSON, Closed geodesics on non-compact Riemannian manifolds, *Math. Z.*, 159, 249-258, (1978).

MULTIBUMP SOLUTIONS AND TOPOLOGICAL ENTROPY.

ERIC SÉRÉ

Ceremade, Université Paris-Dauphine
Place de Lattre de Tassigny, 75775 Paris Cedex 16, France.

Abstract: We present a result on the topological entropy of a large class of Hamiltonian systems. This result is obtained variationally by the construction of "multibump" homoclinic solutions.

§I. INTRODUCTION.

Homoclinic orbits were first introduced by H. Poincaré (see [M][8] for a modern exposition).Considering a hyperbolic fixed point p of a diffeomorphism ϕ, we say that a point $r \neq p$ is homoclinic if it belongs to the intersection of the unstable and stable manifolds W^u, W^s associated to (p, ϕ). The orbit of r is called a homoclinic orbit. Assuming that W^u, W^s intersect transversally at r, and that ϕ is symplectic, Poincaré proved that there are infinitely many homoclinic orbits, geometrically distinct in the following sense:

(the orbits of r, r' are geometrically distinct) $\Leftrightarrow (\forall n \in \mathbb{Z} : \phi^n(r) \neq r')$.

Birkhoff, Smale and other authors also studied homoclinic orbits, and their relation with "Bernoulli shifts". Let us give a precise formulation of a result of Smale (see [M][8]): if $r \neq p$ is a point of transverse intersection of W^u, W^s, then there are $l \in \mathbb{N}^*$, a set I invariant for ϕ^l, and a homeomorphism $\tau : \{0,1\}^{\mathbb{Z}} \longrightarrow I$, such that $\phi^l \circ \tau = \tau \circ \sigma$.
Here, $\sigma((a_n)) = (b_n)$ with $b_n = a_{n+1}$. $\{0,1\}^{\mathbb{Z}}$ is endowed with the standard metric $d(a,b) = \frac{1}{3}\sum_{n \in \mathbb{Z}} \frac{|b_n - a_n|}{2^{|n|}}$. This structure is called a Bernoulli shift.

Bernoulli shifts are an important tool in the study of chaotic behavior. For instance, Smale's result given above implies that the topological entropy of ϕ is greater than $\frac{Ln\ 2}{l}$. This is a direct consequence of the definition of the entropy (see [O][9], p. 182-183):

$$h_{top}(\phi) = \sup_{R>0} \left(\lim_{e \to 0} \left(\limsup_{n \to \infty} \left(n^{-1} Ln\ s(n, e, R) \right) \right) \right),$$

where $s(n, e, R) = \max\{Card(E) : E \subset B(0, R), (\forall x \neq y \in E)(\exists k \in [\![0, n]\!]) : |\phi^k(x) - \phi^k(y)| \geq e\}$.

The results described above were proved by dynamical systems methods, with a transversality assumption on W^u, W^s. In this paper, we want to discuss the following question:

We assume that ϕ is the time-one map of a Hamiltonian system $x' = J\nabla_x H(t, x)$, H being one-periodic in time. Is it possible to say something about Bernoulli shifts and topological entropy, using a variational method? We will see that this approach has several advantages:

• The existence of a homoclinic point r is not an assumption any more, but follows from global hypotheses on H that we call (hA), (hR).

• The only other hypothesis needed to find a positive topological entropy is that the intersection of W^u, W^s is at most countable: this is a weakening of the classical transversality hypothesis.

§II. THE VARIATIONAL APPROACH.

We consider the system

(1) $\quad x' = JAx + J\nabla_x R(t, x) \quad , \quad x \in \mathbb{R}^{2N}, \ t \in \mathbb{R}, \ J = \begin{pmatrix} 0 & -1 \\ 1 & 0 \end{pmatrix}.$

We are looking for non-zero solutions satisfying $x(\pm\infty) = 0$, i.e. solutions homoclinic to 0.
We make the following assumptions:

(hA) $\quad A^* = A$, and $JA = E$ is a constant matrix,
all eigenvalues of which have a non-zero real part.

(hR) $\begin{cases} \bullet \ R(\cdot+1,\cdot) = R(\cdot,\cdot)\text{, and } R \text{ is } C^2. \\ \bullet \ (\forall\, t \in \mathbb{R})\ R(t,\cdot) \text{ is strictly convex.} \\ \bullet \ \text{for some } \alpha > 2\,,\ 0 < k_1 < k_2 < +\infty\,,\text{ we have} \\ \quad \forall\,(t,x) \in \mathbb{R} \times \mathbb{R}^{2N}\,,\quad R(t,x) \leq \dfrac{1}{\alpha}(\nabla_x R, x)\,, \\ \quad k_1|x|^\alpha \leq R(t,x) \leq k_2|x|^\alpha\,. \end{cases}$

Under those assumptions, it was proved in [**CZ-E-S**][3] that there are at least two homoclinic orbits x, y, geometrically distinct, i.e. such that $\forall n \in \mathbb{Z}: n*x \neq y$, where $n*x(t) = x(t-n)$. One of them was obtained by a mountain-pass argument on a dual action functional. This paper has motivated some related work. By a linking argument, Hofer and Wysocki [**H-W**][7] proved the existence of at least one solution to (1) without the convexity assumption. Tanaka [**T**][16] gave another proof of their result, using a method of subharmonics introduced by Rabinowitz [**R**][13].

In the paper [**S**][14], a novel variational argument was introduced, and the following multiplicity result was given:

THEOREM I. *Assume (hA), (hR) are true. Then there are infinitely many orbits homoclinic to 0, geometrically distinct in the sense $x_1 \neq x_2 \iff (\forall n : n*x_1 \neq x_2)$.*

The idea in [**S**][14] was to look for solutions near $(-n)*x + n*x$, where x is the solution found in [**CZ-E-S**][3] by mountain-pass, and n is large enough. We call those solutions "solutions with two bumps". From classical dynamical systems examples, one can guess that such solutions exist for a generic non-autonomous R, but not for an autonomous R. But in the autonomous situation, we have a continuum of solutions which are the translates of x in time, and Theorem I is not contradicted. To give more details, we need some notations.

f is the dual action functional introduced in [**CZ-E-S**][3]. It is defined on the space $L^\beta(\mathbb{R}, \mathbb{R}^{2N})$, with $\frac{1}{\alpha} + \frac{1}{\beta} = 1$. (the exact form of f will be given in §III). $f^a = \{x / f(x) \leq a\}$, $\mathcal{C}$ is the set of non-zero critical points, and $\mathbb{Z}$ acts by integer translations in time.

c is the mountain-pass level, let us define it precisely.

0 is a strict local minimum for f, and $f(0) = 0$. Moreover, f is not bounded from below (see [**CZ-E-S**][3]). So we introduce

$$\Gamma = \{\gamma \in C^0([0,1], L^\beta) \,/\, \gamma(0) = 0\,,\ f \circ \gamma(1) < 0\}\,.$$

Γ is non-empty, and we choose $c = \inf_{\gamma \in \Gamma}(\max f \circ \gamma) > 0$ as mountain-pass level.

In [S][14], the variational gluing of two bumps was possible under the following assumption:

(*): There is some $c' > c$ such that $(\mathcal{C} \cap f^{c'})/\mathbb{Z}$ is finite.

The following result, which is a more precise version of Theorem I, is an immediate consequence of the arguments given in [S][14]:

THEOREM I'. *Assume that (hA), (hR) and (*) are true. Then there are two critical points u, v such that for any $r, h > 0$ and $n \geq N(r, h)$, exists a critical point u_n, with $\|u_n - (-n) * u + n * v\|_{L^\beta} < r$ and $f(u_n) \in [2c - h, 2c + h]$.*

u, v, *possibly equal, satisfy $f(u) = f(v) = c$. u_n is called a solution with two bumps distant of 2n.*

Theorem I is trivial when (*) is not satisfied ("degenerate" situation), and Theorem I' implies Theorem I when (*) is satisfied ("non-degenerate" situation).

In the later work [CZ-R][4], Coti Zelati and Rabinowitz apply the ideas of [S][14] to the case of second order systems, and construct, under assumption (*), solutions with m bumps, i.e. in a ball of center $p^1 * x_1 + ... + p^m * x_m$ and radius ε, for the norm $W^{1,2}(\mathbb{R}, \mathbb{R}^N)$. The x_i are in a fixed finite set of critical points of the action functional $\int \frac{\dot{x}^2}{2} - V$ found thanks to a montain-pass, and for any i, $(p^{i+1} - p^i) \geq K(\varepsilon, m)$. In the construction of [CZ-R][4], the minimal distance K between bumps goes to infinity as m goes to infinity, for ε fixed.

Other applications, in the domain of partial differential equations, are given in [CZ-R][5], [LI][10], [LI][11].

In the paper [C-L][2] of Chang and Liu, the assumption (*) is replaced by

(**): $\mathcal{C} \cap f^{c'}$ contains only isolated points.

In the recent work [S][15], (**) is replaced by a weaker assumption

$(\mathcal{H}) : \mathcal{C} \cap f^{c'}$ is at most countable.

Moreover, multibump solutions are constructed for a minimal distance K between bumps independent of m. This last point, whose proof requires many modifications in the arguments of [S][14], [CZ-R][4], allows to study the topological entropy of the Hamiltonian system. The main theorem in [S][15] can be stated as follows:

THEOREM II.

Assume (hA), (hR) and $\mathcal{H}$ are true. Then there exists a homoclinic orbit x such that, for any $\varepsilon > 0$, and any finite sequence of integers $\overline{p} = (p^1, ..., p^m)$, satisfying

$$(\forall i) : (p^{i+1} - p^i) \geq K(\varepsilon),$$

there is a homoclinic orbit $y_{\overline{p}}$, with

$$(\forall t \in \mathbb{R}) : |y_{\overline{p}}(t) - \sum_{i=1}^{m} x(t - p^i)| \leq \varepsilon.$$

Here, K is a constant independent of m.

The assumption $(\mathcal{H})$ cannot be satisfied in the autonomous situation, where the translates of x in time form a continuum. Now, if W^u, W^s intersect transversally, then their intersection is at most countable, and so is the set of homoclinic solutions; but the converse is false.

Since K does not depend on m, we can study the limit $m \to \infty$ (see [S][15]), and get solutions with infinitely many bumps (those are not homoclinic orbits any more). We have

COROLLARY II.1.

With the hypotheses and notations of Theorem II, for any interval $I \subset \mathbb{Z}$, finite or infinite, and any sequence of integers $\overline{p} = (p^i)_{i \in I}$ such that $(\forall i) : (p^{i+1} - p^i) \geq K(\varepsilon)$, there is a solution $y_{\overline{p}}$ of (1) satisfying

$$(\forall t \in \mathbb{R}) : |y_{\overline{p}}(t) - \sum_{i \in I} x(t - p^i)| \leq \varepsilon.$$

If I is infinite, we say that y has infinitely many bumps.

As a consequence, we have an "approximate" Bernoulli shift structure:

COROLLARY II.2.

Under the hypotheses of Theorem II, there is $x_0 \in \mathbf{R}^{2N} \setminus \{0\}$ such that, for any $\varepsilon > 0$, exist $K = K(\varepsilon) > 0$ and $\tilde{\tau} = \tilde{\tau}(\varepsilon) : (\{0,1\}^{\mathbf{Z}}, d) \to (\mathbf{R}^{2N}, |\cdot|)$, with:

- $\tilde{\tau}$ is injective, and $\tilde{\tau}^{-1}$ is uniformly continuous.
- $(\forall\, n \in \mathbf{Z}) \quad \|\tilde{\tau} \circ \sigma^n - \phi^{Kn} \circ \tilde{\tau}\|_\infty < 2\varepsilon$.
- $\begin{cases} s_0 = 1 \Rightarrow |\tilde{\tau}(s) - x_0| < \varepsilon \\ s_0 = 0 \Rightarrow |\tilde{\tau}(s)| < \varepsilon\,. \end{cases}$

Here, ϕ is the time-one flow of (1), and $\sigma(s)_n = s_{n+1}$. Note that we cannot say that $\tilde{\tau}$ is continuous. We call $(\tilde{\tau}(\{0,1\}^{\mathbf{Z}}), \phi^K)$ an approximate Bernoulli shift structure.

To prove Corollary II.2, just take $x_0 = x(0)$, and $\tilde{\tau}(s) = y_{\overline{p}}(0)$, where $p^i = K\, q^i$, $s_n = \chi_{\{q^i; i \in I\}}(n)$. We refer to [S][15] for more details.

Now, we are in a position to state the result on topological entropy. Choose $\varepsilon \leq \frac{|x_0|}{3}$. If two sequences s, s' are such that $s_k \neq s'_k$ for some k, then
$$\left|\Phi^{K(\varepsilon)k} \circ \tau(s) - \Phi^{K(\varepsilon)k} \circ \tau(s')\right| \geq \frac{|x_0|}{3}\,.$$

So, for $e < \frac{|x_0|}{3}$ and $R > |x_0| + \varepsilon$, we get $s(Kn, e, R) \geq 2^n$, and $h_{top}(\phi) \geq \frac{\mathrm{Ln}\, 2}{K(\varepsilon)}$. So Corollary II.2 implies

COROLLARY II.3.

With the hypotheses of Theorem I, the flow of (1) has a positive topological entropy.

Independently of [S][15], Bessi in [B][1] constructs variationally an approximate Bernoulli shift for the one-dimensional pendulum. He replaces assumption (*) by a weakening of the classical Melnikov condition, and his result is given for small perturbations of an autonomous system.

§III. SKETCH OF THE PROOF OF THEOREM II.

We use a variational formulation based on Clarke's dual action principle (see [E][6], [CZ-E-S][3]). Define $G(t,y) = \max\{(z \cdot y) - R(t,z) \ / \ z \in \mathbf{R}^{2N}\}$. G is 1-periodic in time, strictly convex in y, and satisfies, for $\frac{1}{\alpha} + \frac{1}{\beta} = 1$:

$$\begin{cases} 0 \leq \frac{1}{\beta}(\nabla_y G, y) \leq G(t,y) \leq (\nabla_y G, y) \\ (\exists \ c_1, c_2 > 0) \ (\forall \ (y,t)) \ : \quad c_1 |y|^\beta \leq G(t,y) \leq c_2 |y|^\beta, \\ \qquad |\nabla_y G(t,y)| \leq c_2 |y|^{\beta-1}. \end{cases}$$

Define
$$D : W^{1,\beta}(\mathbf{R}, \mathbf{R}^{2N}) \rightarrow L^\beta(\mathbf{R}, \mathbf{R}^{2N})$$
$$z \mapsto (-J\frac{d}{dt} - A)z,$$
$$L = D^{-1}.$$

We call $\mathcal{C}$ the set of non-zero critical points of the following functional f :

$$f(u) = \int G(t,u) \, dt - \frac{1}{2} \int (u, Lu) \, dt \quad, \quad u \in L^\beta(\mathbf{R}, \mathbf{R}^{2N}).$$

We have (see [CZ-E-S][3])

LEMMA 1. *If $u \in \mathcal{C}$, then $x = Lu$ is a non-zero solution of (1) such that $x(\pm\infty) = 0$, i.e. an orbit homoclinic to 0.*

Our task will be to find a large class of elements of $\mathcal{C}$.

For this purpose, we need some compactness properties of f. Unfortunately, f does not satisfy the Palais-Smale (PS) condition, because it is invariant for the action of the non-compact group $\mathbf{Z} : n * u = u(\cdot - n)$. To deal with this problem, we use the concentration-compactness theory of P.L. Lions (see [LS][12], [CZ-E-S][3]). We have:

LEMMA 2. *Suppose (hA), (hR) are true. Then f satisfies the following compactness property:*

Let $(u_n)_{n \geq 0}$ be a sequence such that

$$f(u_n) \text{ is bounded}, \quad f'(u_n) \rightarrow 0.$$

Then there exist $m \in \mathbf{N}$, a subsequence $(n_p)_{p \geq 0}$, and $u^1, \ldots, u^m$ in $\mathcal{C}$, not necessarily distinct, such that

$$\left\| u_{n_p} - \sum_{i=1}^{m} k_p^i * u^i \right\| \xrightarrow[p \to \infty]{} 0,$$

where $k_p^i \in \mathbb{Z}$, $(k_p^j - k_p^i) \to +\infty$ as $p \to +\infty$ if $i < j$.

Note that the sums $\sum_{i=1}^{m} k^i * u^i$ do not vary continuously when $k^1 ... k^m$ vary in $\mathbb{Z}$. This leads to introduce a new compactness condition (see [**CZ-E-S**]³, [**S**]¹⁴,¹⁵). We call $\overline{PS}$ sequence a sequence (u_n) such that $f(u_n)$ is bounded, $f'(u_n) \to 0$, and $(u_{n+1} - u_n) \to 0$. We have

LEMMA 3. *Assume* (hA), (hR) *and* ($\mathcal{H}$) *are true. Let* (u_n) *be a* $\overline{PS}$ *sequence such that* $f(u_n) \leq c'$. *Then* (u_n) *is convergent. We say that* f *satisfies the* $\overline{PS}$ *condition under level* c'.

The interest of $\overline{PS}$ is that, if f is bounded on a pseudo-gradient line, then one can find a $\overline{PS}$ sequence on this line. So $\overline{PS}$ can give the same kind of deformation lemmas as the Palais-Smale condition. If $\overline{PS}$ is satisfied under level c', by deforming a particular curve in Γ, one finds at least one critical point u between levels c and c'. When (*) holds, one can impose $f(u) = c$. When the weaker assumption ($\mathcal{H}$) holds, the best that can be done is to take u with $f(u) - c$ arbitrarily small.

In [**S**]¹⁴, under assumption (*), a "product min-max" is constructed at level 2c, for the "split" functional $\tilde{f}(x) = f(x\chi_{\mathbb{R}_-}) + f(x\chi_{\mathbb{R}_+})$, where χ_I is the caracteristic function of I. Theorems I and I' are then proved by contradiction, thanks to a deformation argument. This argument works because the differentials f' and $\tilde{f}'$ "look the same" near $(-n) * u + n * v$, where u, v are critical points associated to the mountain-pass, possibly equal.

The proof of Theorem II given in [**S**]¹⁵ is based on the same ideas, but contains several technical improvements.

We first construct, for any r, $h > 0$, a non-trivial homology class in $H_1(f^{\bar{c}+h}, f^{\bar{c}-h})$, containing a chain included in $B(u,r)$, thanks to assumption $\mathcal{H}$. Here, $\bar{c} = f(u) \in [c, c']$, and $u \in \mathcal{C}$, found thanks to the mountain-pass, is independent of r, h.

Then, roughly speaking, we consider a product of m "copies" of this homology class, and find a "product min-max" in a neighborhood of $\sum_{i=1}^{m} p^i * u$. This is done thanks to Künneth's formula, $H_*(X \times Y, (Z \times Y) \cup (X \times T)) = H_*(X, Z) \otimes H_*(Y, T)$. Note that in [S][14], [CZ-R][4], a more elementary procedure (without homology) is used to construct the product min-max. It would be possible to use this procedure in the proof of Theorem II. But the use of homology allows an easy generalization to situations where u would not be found by mountain-pass, but by a min-max on a family of sets of higher dimension.

Finally, we find a critical point $u_{\bar{p}}$ in a neighborhood of $\sum_{i=1}^{m} p^i * u$, provided $(p^{i+1} - p^i) \geq K$, K depending only on r, not on m. To do this, we assume that $u_{\bar{p}}$ does not exist, construct a more precise version of the deformation used in [S][14], and apply it to the "product min-max" to obtain a contradiction.

In the proof of Theorem II, a crucial point is to make a suitable choice of the neighborhood of $\sum_{i=1}^{m} p^i * u$ in which we want to find $u_{\bar{p}}$: with the choice made in [CZ-R][4], i.e. balls of fixed radius for a norm independent of m, it seems impossible to control K as m increases. To define the correct neighborhood, we have to introduce some notations.

Take $x \in L^\beta$, $\bar{p} = (p^1, \ldots, p^m) \in \mathbb{Z}^m$, $m \geq 1$, $p^i < p^{i+1}$. Denote

$$x_i = x\chi_{\left[\frac{p^{i-1}+p^i}{2}, \frac{p^i+p^{i+1}}{2}\right]}, \quad f_i(x) = f(x_i),$$

with χ_I the characteristic function of I, $p^0 = -\infty$, $p^{m+1} = +\infty$.
We have $x = \sum_{i=1}^{m} x_i$, but $f \neq \sum_{i=1}^{m} f_i$.

Consider the sets

$$\mathcal{L}_+(h) = \bigcap_{i=1}^{m} (f_i)^{\bar{c}+h} \quad , \quad \mathcal{L}_-(h) = \mathcal{L}_+(h) \cap \bigcup_{i=1}^{m} (f_i)^{\bar{c}-h},$$

and the "product" ball

$$B^u_{\bar{p}, \rho} = \left\{ x \in L^\beta \;/\; (\forall\, i)\, \|(x - p^i * u)_i\|_{L^\beta} < \rho \right\}$$

for $\rho > 0$, $u \in \mathcal{C}$.

We choose the neighborhood $U_{\bar{p}, r, h} = B^u_{\bar{p}, r} \cap \left(\mathcal{L}_+(h) \setminus \mathcal{L}_-(h) \right)$.
We have (see [S][15]):

THEOREM III. *Assume ($\mathcal{H}$) is true.*
Then there is $u \in \mathcal{C}$, with $f(u) = \bar{c} \in [c, c'[$, and such that for any $r, h > 0$, for all $m \geq 1$ and $\bar{p} = (p^1, .., p^m) \in \mathbb{Z}^m$:

$$\left[(\forall\, i) \,:\, (p^{i+1} - p^i) \geq K(r, h)\right] \;\Rightarrow\; \left[\mathcal{C} \cap U_{\bar{p}, r, h} \neq \emptyset\right] .$$

$K(r, h)$ *is independent of m.*

Keeping h fixed, Theorem II is a direct consequence of Theorem III.

The method of proof described above can be applied to other functionals. For instance, one can prove a result analogous to Theorem II for the equations of [CZ-R][4,5] (details will be given in a forthcoming paper). It seems more difficult to extend the result to the non-convex Hamiltonian systems studied in [H-W][7], [T][16], because in this case, the functional is strongly indefinite, and one expects technical complications in the product min-max procedure.

Bibliography.

[B][1] U. Bessi, *A variational proof of a Sitnikov-like theorem*, preprint Scuola Normale Superiore.

[C-L][2] K.C. Chang and J.Q. Liu, *A remark on the homoclinic orbits for Hamiltonian systems*, research report of Peking University.

[CZ-E-S][3] V. Coti-Zelati, I. Ekeland, E. Séré, *A variational approach to homoclinic orbits in Hamiltonian systems*, Mathematische Annalen, 288, (1990), 133-160.

[CZ-R][4] V. Coti-Zelati and P. Rabinowitz, *Homoclinic orbits for second order Hamiltonian systems possessing superquadratic potentials,*, preprint Sissa.

[CZ-R][5] V. Coti-Zelati and P. Rabinowitz, *Homoclinic type solutions for a semilinear elliptic PDE on $\mathbb{R}^n$*, preprint Sissa.

[E][6] I. Ekeland, *Convexity Methods in Hamiltonian Systems*, Springer Verlag, 1989.

[**H-W**][7] H. Hofer and K. Wysocki, *First order elliptic systems and the existence of homoclinic orbits in Hamiltonian systems*, Math. Annalen 288 (1990), 483-503.

[**M**][8] J. Moser, *Stable and random motions in Dynamical Systems*, Princeton University Press, Princeton, 1973.

[**O**][9] Séminaire Orsay, *Travaux de Thurston sur les surfaces*, Astérisque 66-67, Société Mathématique de France.

[**LI**][10] Y.Y. Li, *On $-\Delta u = k(x)u^5$ in $\mathbb{R}^3$*, preprint Rutgers University.

[**LI**][11] Y.Y. Li, *On prescribing scalar curvature problem on S^3 and S^4*, preprint Rutgers University.

[**LS**][12] P.L. Lions, *The concentration-compactness principle in the Calculus of Variations*, Revista Iberoamericana 1, (1985), 145-201.

[**R**][13] P. Rabinowitz, *Homoclinic orbits for a class of Hamiltonian systems*, Proc. Roy. Soc. Edinburgh, 114 A, (1990), 33-38.

[**S**][14] E. Séré, *Existence of infinitely many homoclinic orbits in Hamiltonian systems*, Math. Zeitschrift 209, (1992), 27-42.

[**S**][15] E. Séré, *Looking for the Bernoulli shift*, preprint CEREMADE.

[**T**][16] K. Tanaka, *Homoclinic orbits in a first order superquadratic Hamiltonian system : convergence of subharmonics*, preprint Nagoya University.

[**W**][17] S. Wiggins, *Global Bifurcations and Chaos*, Applied Mathematical Sciences 73, Springer-Verlag.

AVOIDING COLLISIONS IN SINGULAR POTENTIAL PROBLEMS

ENRICO SERRA

Dipartimento di Matematica, Università di Torino
V. Carlo Alberto 10, 10123 Torino, Italy

1. INTRODUCTION

In this note we shall describe some results obtained in the last few years concerning a classical problem encountered in the study of dynamical systems with singular potentials. We shall present some well–known facts and at the same time we shall give some applications of recent results to various kinds of problems.

We recall the abstract setting of singular potential problems.

Let $\Omega \subset \mathbf{R}^N$ ($N \geq 2$) be an open set and let $V \in \mathcal{C}^1(\mathbf{R} \times \Omega; \mathbf{R})$. We look for solutions to the system of differential equations

$$\ddot{q}(t) + V'(t, q(t)) = 0 \tag{1.1}$$

where $q(t) \in \mathbf{R}^N$ for all t and V' denotes the gradient of V with respect to the second variable.

This system can be coupled with various kinds of boundary conditions, which give rise to different types of solutions. Concerning periodic solutions we have the fixed period problem, consisting in looking for solutions such that $q(t+T) = q(t)$ for all t (T is usually the period of V in the first variable) and, when V does not depend on t, the fixed energy problem, dealing with periodic solutions (the period is unknown) on the energy hypersurface

$$\frac{1}{2}|\dot{q}|^2 + V(q) = h$$

where $h \in \mathbf{R}$ is a prescribed number.

In the last few years there has also been a considerable interest in *homoclinic* (and *heteroclinic*) solutions, that is, solutions to (1.1) defined for $t \in \mathbf{R}$ and approaching as $t \to \pm\infty$ one (or two) critical points of the potential V, with velocity $\dot{q}(t) \to 0$ as $t \to \pm\infty$.

In this kind of problems the potential is called "singular" when it is assumed that

$$\lim_{x \to \partial\Omega} V(t,x) = -\infty \qquad \text{(uniformly in } t\text{)}. \tag{1.2}$$

The mechanical interpretation of (1.1) and (1.2) is the description of the motion of a particle located in $q(t) \in \mathbf{R}^N$ and moving subject the force $V'(t, q(t))$; condition (1.2) says that the boundary of Ω attracts the particle.

In order to fix ideas, we shall always think of Newton's law of gravitation, which fits into the above scheme when $\Omega = \mathbf{R}^N \setminus \{0\}$ and $V(t, x) = -\frac{a}{|x|}$, $(a > 0)$.

In this note we are interested in solutions to (1.1) when the potential V behaves like $-\frac{a}{|x|^\alpha}$, for some $a, \alpha > 0$ near the origin, so that we assume, from now on, $\Omega = \mathbf{R}^N \setminus \{0\}$.

Singular potential problems are thus generalizations of the classical two–body problem. Slight variants of this abstract setting also permit to deal with more general N–body systems.

2. THE VARIATIONAL APPROACH

The first results appeared in the literature concerned the fixed period problem, which we now take as the model case. Although classical, this problem has recently received powerful contributions, mainly due to the new approach based on variational methods. Indeed periodic (but also homoclinic and heteroclinic) solutions to (1.1) can be found as critical points of suitable functionals. In the case of the fixed period problem, for instance, it is readily seen that T–periodic solutions to (1.1) correspond to critical points of the action functional

$$I(q) = \frac{1}{2} \int_0^T |\dot{q}|^2 dt - \int_0^T V(t, q) dt, \tag{2.1}$$

in the set $\Lambda = \{q \in H^1(S^1; \mathbf{R}^N) \,/\, q(t) \neq 0 \ \forall t\}$.

The search for critical points of functionals of this type presents, even in simple model cases such as $V(t,q) = -\frac{a}{|q|^\alpha}$, $(a, \alpha > 0)$, two main dificulties.

The first one originates from the fact that the natural assumptions on V at infinity are

$$V(t,x) \to 0 \quad \text{and} \quad V'(t,x) \to 0 \quad \text{as} \quad |x| \to +\infty; \qquad (2.2)$$

these assumptions give rise to the typical *lack of compactness* present in many variational problems. Indeed if (2.2) holds then the functional I does not satisfy the Palais–Smale condition (PS for short hereafter), for instance at level zero. In order to overcome this inconvenience various techniques have been developed, for which we refer to [2]–[4], [6], [13]–[15], [24], [26], [27].

Here we are mainly concerned with a second type of difficulty which is peculiar of singular potential problems: the *lack of completeness*. To illustrate this, suppose that $(q_n)_{n \in \mathbf{N}} \subset \Lambda$ is a PS sequence at some level c, namely, $I(q_n) \to c$ and $I'(q_n) \to 0$ as $n \to \infty$. If the PS condition holds at level c, then, along a subsequence, $q_n \to q \in \overline{\Lambda}$, but it may happen that $q \in \partial \Lambda = \{q \in H^1(S^1; \mathbf{R}^N) \,/\, \exists t_0 \in S^1, \, q(t_0) = 0\}$. The set $\partial \Lambda$ is the set of *collision* orbits. In other words, standard variational arguments cannot be directly applied because the natural setting of the problem is the set Λ, which is open (in $H^1(S^1; \mathbf{R}^N)$). From the mechanical point of view a periodic solution $q \in \partial \Lambda$, that is a "collision" periodic solution, is meaningless. One therefore needs supplementary arguments to avoid the presence of collision solutions.

In general these arguments are structured as follows. First one tries to locate the levels at which collision solutions are not allowed, and then one builds a variational argument (such as a minimax approach), working at these safe levels, where it is also hoped that the PS condition holds.

The simplest assumption used to locate the collision levels is the so-called "Strong Force" condition, introduced by Gordon in [14]:

$$\exists \varepsilon > 0, \, \exists U \in \mathcal{C}^1(B_\varepsilon \setminus \{0\}; \mathbf{R}) \text{ such that } \lim_{x \to 0} U(x) = \infty \text{ and} \qquad (SF)$$
$$-V(t,x) \geq |U'(x)|^2 \quad \forall x \in B_\varepsilon \setminus \{0\}, \, \forall t.$$

A potential V satisfying (SF) is called a strong force. The power of this condition is the fact that if V is a strong force, then

$$\liminf_{q \to \partial \Lambda} I(q) = +\infty, \qquad (2.3)$$

so that no finite level of the action functional can contain a collision orbit. Therefore any variational argument working at a finite level, as is usually the case, yields (modulo PS) a noncollision solution to (1.1). The main drawback in the use of (SF) is the fact that if V behaves like $-\frac{a}{|x|^\alpha}$ near zero, then necessarily $\alpha \geq 2$, which means that the Kepler problem and its perturbations (which originally motivated this kind of studies) cannot be treated under (SF). For results concerning potentials satisfying (SF) we refer to [2], [6], [15] and references therein.

After those papers a considerable effort has been devoted to the search of weaker conditions than (SF), and in particular suited to deal with the cases $V(t,x) \approx -\frac{a}{|x|^\alpha}$, with $\alpha \in]0,2[$. The first success in this context is due to Degiovanni and Giannoni ([13]), who introduced a "pinching inequality" in the study of (1.1). Roughly speaking, suppose that V satisfies

$$\frac{a}{|x|^\alpha} \leq -V(x) \leq \frac{b}{|x|^\alpha} \qquad \forall x \neq 0, \tag{2.4}$$

with $b - a$ small, depending on α. Then, if $\alpha > 1$, one can find a minimax class $\Gamma \subset 2^\Lambda$ such that

$$c := \inf_{A \in \Gamma} \max_{q \in A} I(q) < \inf_{q \in \partial \Lambda} I(q), \tag{2.5}$$

so that no collision are allowed at level c. This type of condition has been used repeatedly in the literature, see e.g. [8], [13] and [26], [27], where pinching inequalities on V' are introduced to obtain multiple solutions both for the fixed period problem and for the fixed energy problem. Remark that the case $\alpha = 1$ is still not covered by these assumptions.

Finally we wish to recall a result by Ambrosetti and Coti Zelati, of a somewhat different nature. In [3] indeed, a bifurcation argument is used to prove that multiple solutions arise for small, symmetric perturbations of the Kepler problem.

These are thus the main types of contitions which have been used in the study of singular potential problems. Minor modification of the above outlined arguments have been introduced to obtain a series of corresponding results in the case of the fixed energy problem. Lastly, the only results concerning homoclinics for singular potential problems are those of [1], [8] and [25], where (SF) is used.

3. LOCAL MINIMA AND COLLISIONS

The need of new type of conditions to replace (SF) and pinching motivated the results contained in [22] and the subsequent papers [11], [12], [21], [23], which extended the original results to other kind of problems.

The starting point is the observation that since the collision is a phenomenon of a somewhat *local* nature, it might be avoided using only local hypotheses near the singularity of the potential (in contrast to assumptions of the type (2.4)) and possibly much weaker than (SF), including potentials behaving like $-\frac{a}{|x|^\alpha}$, for any $\alpha \in]0,2[$. The key result is given in Theorem 3.1 below; in order to state it we introduce the following setting.

Let $T > 0$, $V \in C^2([0,T] \times \mathbf{R}^N \setminus \{0\}; \mathbf{R})$ and write V in the form

$$V(t,x) = -\frac{a}{|x|^\alpha} + U(t,x) \tag{V1}$$

for some $a > 0$ and $\alpha \in]0,2[$. Given two points $x_0, x_1 \in \mathbf{R}^N \setminus \{0\}$ let

$$E = \{u \in H^1([0,T]; \mathbf{R}^N) \ / \ u(0) = x_0, \ u(T) = x_1\} \tag{3.1}$$

and let $f : E \to \mathbf{R}$ be the functional

$$f(u) = \begin{cases} \frac{1}{2}\int_0^T |\dot{u}|^2 dt - \int_0^T V(t,u)dt & \text{if this quantity is finite} \\ +\infty & \text{otherwise} \end{cases} \tag{3.2}$$

Then the following result holds.

THEOREM 3.1. ([22]). Let V be as in (V1), with $N \geq 3$, $a > 0$ and $\alpha \in]0,2[$. Assume V satisfies

$$\limsup_{x \to 0} \left|\frac{\partial U}{\partial t}(t,x)\right| |x|^\alpha < +\infty, \quad \text{uniformly in } t \tag{V2}$$

and

$$\exists M > 0, \ \exists \sigma > 0 \quad \text{such that}$$
$$\limsup_{x \to 0} |V''(t,x)||x|^{\alpha+2-\sigma} \leq M, \quad \text{uniformly in } t. \tag{V3}$$

If $q \in E$ is a local minimum for f on E, then

$$q(t) \neq 0 \quad \forall t \in [0,T].$$

REMARK 3.2. The main difference between this result and the ones quoted above is in the fact that Theorem 3.1 involves only a property of special orbits, and not a property of the entire level sets of f. Indeed, Theorem 3.1 states that at every level, the properties $q(t) = 0$ for some t and q local minimum for f are incompatible. Thus the information that q is a local minimum for the action functional, at any level, implies that q is a noncollision orbit.

Clearly every function $V \in C^2([0,T] \times \mathbf{R}^N \setminus \{0\}; \mathbf{R})$ can be written in the form (V1). The only assumptions in the theorem are thus (V2) and

(V3). These permit to deal with singular perturbations of $-\frac{a}{|x|^\alpha}$, but if U is everywhere regular, (V2) and (V3) are trivially satisfied, so that we obtain the following result.

COROLLARY 3.3. *Let V be as in (V1), with $N \geq 3$, $a > 0$ and $\alpha \in]0,2[$. Assume $U \in C^2([0,T] \times \mathbf{R}^N; \mathbf{R})$. If $q \in E$ is a local minimum for f on E, then*
$$q(t) \neq 0 \quad \forall t \in [0,T].$$

Theorem 3.1 and Corollary 3.3 do not assert the existence of minimizers of f on E. Conditions which yield existence results for various kinds of problems are given in the next section.

SKETCH OF THE PROOF OF THEOREM 3.1. The proof of Theorem 3.1 under hypotheses (V2) and (V3) is quite long and technical, but it can be considerably simplified if we strenghten a little the assumptions. Indeed we suppose from now on that the function U in (V1) is locally radially symmetric in a small ball around $x = 0$. By this we mean that U satisfies

$$\exists r > 0 \ \exists \phi \in C^2(]0,r]; \mathbf{R}) \quad \text{such that} \quad (V4)$$
$$U(t,x) = \phi(|x|) \quad \forall 0 < |x| \leq r, \quad \forall t.$$

We now sketch the proof under assumptions (V2)–(V4).

Suppose that $q \in E$ is a local minimizer of f on E, and assume for contradiction that $q(\bar{t}) = 0$, for some $\bar{t} \in]0, T[$. Since q is a local minimizer, in every connected component S of $[0,T] \setminus \{t \ / \ q(t) = 0\}$ it satisfies the Euler–Lagrange equations associated to f, which, when $|q(t)| \leq r$ read

$$-\ddot{q}(t) = a\alpha \frac{q(t)}{|q(t)|^{\alpha+2}} + \phi'(|q(t)|) \frac{q(t)}{|q(t)|}, \tag{3.3}$$

and the energy conservation, given by

$$\frac{1}{2}|\dot{q}(t)|^2 - \frac{a}{|q(t)|^\alpha} + \phi(|q(t)|) = E_S, \tag{3.4}$$

for some constant $E_S \in \mathbf{R}$. From (3.3), (3.4) and (V3) we see that

$$\frac{d^2}{dt^2}|q(t)|^2 > 0$$

for all t near a time of collision. This means in particular that collisions are isolated and we can assume that there exists $\delta > 0$ such that

i) q is regular in $\mathcal{T} =]\bar{t} - \delta, \bar{t}[\cup]\bar{t}, \bar{t} + \delta[$,
ii) $|q(t)| \le r \quad \forall t \in \mathcal{T}$.

Since in B_r the potential is radially symmetric, $\frac{q(t)}{|q(t)|}$ is constant both in $]\bar{t} - \delta, \bar{t}[$ and in $]\bar{t}, \bar{t} + \delta[$; in particular the orbit q lies on a fixed plane for all $t \in \mathcal{T}$.

Therefore there exists $v \in S^{N-1}$ such that $v \cdot q(t) = 0 \ \forall t \in \mathcal{T}$. We now define a piecewise linear function $w : [0, T] \to \mathbf{R}^N$ as

$$w(t) = \begin{cases} 0 & \text{if } t \notin \mathcal{T} \\ \mu v & \text{if } t \in [\bar{t} - \frac{\delta}{2}, \bar{t} + \frac{\delta}{2}] \end{cases}$$

where μ will be conveniently chosen later. It is clear that $w(t) \cdot q(t) = \dot{w}(t) \cdot \dot{q}(t) = 0 \ \forall t \in [0, T]$.

We are going to prove that for all μ small enough, $f(q + w) < f(q)$, contradicting the fact that q is a minimum point. To this aim we evaluate

$$f(q + w) - f(q) = \int_{\bar{t}-\delta}^{\bar{t}+\delta} |\dot{w}|^2 dt - \int_{\bar{t}-\delta}^{\bar{t}+\delta} [V(q + w) - V(q)] dt.$$

To begin with, notice that plainly

$$\int_{\bar{t}-\delta}^{\bar{t}+\delta} |\dot{w}|^2 dt \le C_1 \mu^2,$$

for some constant C_1 independent of μ. Next we have

$$\int_{\bar{t}-\delta}^{\bar{t}+\delta} [-V(q+w) + V(q)] dt = \int_{\bar{t}-\delta}^{\bar{t}+\delta} \int_0^1 \frac{d}{d\lambda}(-V(q+\lambda w)) \, d\lambda dt =$$
$$= \int_{\bar{t}-\delta}^{\bar{t}+\delta} \int_0^1 (-V'(q+\lambda w) \cdot w) \, d\lambda dt. \quad (3.5)$$

Now, if δ and μ are small enough, then $|q(t) + \lambda w(t)| < r$, so that by (V3) and the fact that $q(t)$ and $w(t)$ are orthogonal, a direct computation shows

$$V'(q + \lambda w) \cdot w \ge \frac{a\alpha}{2} \frac{\lambda |w|^2}{|q + \lambda w|^{\alpha+2}}$$

Carrying this inequality in (3.5) it is easy to see that

$$\int_{\bar{t}-\delta}^{\bar{t}+\delta} [-V(q+w) + V(q)] dt \le -\frac{a\alpha}{4} \mu^2 \int_{\bar{t}-\frac{\delta}{2}}^{\bar{t}+\frac{\delta}{2}} \int_{\frac{1}{2}}^1 \frac{d\lambda dt}{|q + \lambda w|^{\alpha+2}}.$$

Finally, since $q(\bar{t}) = 0$, we have

$$|q(t)| \leq |\int_{\bar{t}}^{t} |\dot{q}|dt| \leq C_2 |t - \bar{t}|^{\frac{1}{2}},$$

so that $|q(t)| \leq \mu$ whenever $|t - \bar{t}| \leq \sigma_\mu = \frac{\mu^2}{C_2^2}$. Restricting the interval of integration to $[\bar{t} - \sigma_\mu, \bar{t} + \sigma_\mu]$, and noticing that there $|q + \lambda w| \leq 2\mu$, we obtain

$$\int_{\bar{t}-\delta}^{\bar{t}+\delta} [-V(q+w) + V(q)]dt \leq -\frac{a\alpha}{4}\mu^2 \int_{\bar{t}-\sigma_\mu}^{\bar{t}+\sigma_\mu} \int_{\frac{1}{2}}^{1} \frac{d\lambda\, dt}{2^{\alpha+2}\mu^{\alpha+2}} \leq -C_3 \mu^{2-\alpha}.$$

Thus

$$f(q+w) - f(q) \leq C_1 \mu^2 - C_3 \mu^{2-\alpha} < 0$$

if μ is small enough, and a contradiction is reached. ∎

REMARK 3.4. The complete proof of Theorem 3.1 (without assumption (V4)) is reported in [22]. It consists in showing that even if the motion is no longer planar around $x = 0$, it is "almost" planar, due to the fact that under (V2) and (V3) the main term in the potential is radially symmetric. Then, using much sharper estimates than those given above, it is possible to show that the same kind of variation, along an "almost" normal direction makes the functional decrease.

4. SOME APPLICATIONS

We now sketch some of the possible applications of Theorem 3.1. This result applies, as we have seen, whenever we can deal with (local) minimizers of the action functional I. Some problems, however, do not admit any minimizers, or it may happen that the minimum point corresponds to a "trivial" solution. This difficulty can be overcome, in some cases, with the introduction of suitable constraints on the function space. We now present some results in this framework, referring to [12], [21], [22] for more results and compete proofs.

A) THE FIXED PERIOD PROBLEM. The first application comes from [22] and actually motivated Theorem 3.1.

Assume $V \in C^2(\mathbf{R} \times \mathbf{R}^N \setminus \{0\}; \mathbf{R})$ is of the form (V1) and is T-periodic in the first variable. We wish to find T-periodic solutions to the problem

$$\begin{cases} \ddot{q} + V'(t,q) = 0 & \forall t \\ q(t+T) = q(t) & \forall t \\ q(t) \neq 0 & \forall t. \end{cases} \quad (4.1)$$

In the spirit of Theorem 3.1 we require that V satisfies (V2) and (V3), so that collisions can be avoided among the minimizers of the functional $I(q) = \frac{1}{2}\int_0^T |\dot{q}|^2 dt - \int_0^T V(t,q)dt$. In such a general situation, I need not posses any (local) minima over $H^1(S^1; \mathbf{R}^N)$, and therefore some further assumption is in order.

THEOREM 4.1. ([22]). Assume V satisfies (V1)–(V3) with $N \geq 3$, $a > 0$ and $\alpha \in]0, 2[$, and

$$\lim_{|x|\to +\infty} \frac{|V'(t,x)|}{|x|} = 0 \quad \text{uniformly in } t \tag{V4}$$

$$V(t + \frac{T}{2}, -x) = V(t, x) \quad \forall x \neq 0, \ \forall t \tag{S}$$

Then problem (4.1) has at least one solution.

The role of (V4) and (S) in the proof of this result is clear: restricting I to the space of antiperiodic functions ($q(t + \frac{T}{2}) = -q(t)$), I turns out to be coercive (and weakly lower semicontinuous). Therefore there exists a minimum point q. With a slight variant of the proof of Theorem 3.1 (due to the fact that variations must be antiperiodic) it is easy to see that $q(t) \neq 0$ for all t, so that q solves (4.1).

B) HETEROCLINICS AT INFINITY. Suppose $V \in C^2(\mathbf{R}^N \setminus \{x_0\}; \mathbf{R})$ satisfies

$$\exists R > 0, \ \exists U \in C^2(\mathbf{R}^N \setminus B_R; \mathbf{R}) \quad \text{such that}$$
$$i) \ V(x) = -\frac{1}{|x|^\beta} + U(x) \quad \text{for all } |x| \geq R \tag{H1}$$
$$ii) \lim_{|x|\to +\infty} U(x)|x|^\beta = 0,$$

$$V(x) \leq 0 \quad \forall x \in \mathbf{R}^N \setminus \{x_0\}, \quad V(x) = 0 \text{ if and only if } x = 0, \tag{H2}$$

$$\exists \varepsilon > 0 \ \exists W \in C^2(B_\varepsilon(x_0) \setminus \{x_0\}; \mathbf{R}) \quad \text{such that}$$
$$V(x) = -\frac{1}{|x - x_0|^\alpha} + W(x) \quad \forall x \in B_\varepsilon(x_0) \setminus \{x_0\}, \tag{H3}$$

and

$$\exists C > 0 \ \exists \sigma > 0 \quad \text{such that} \quad \limsup_{x \to x_0} |W''(x)||x - x_0|^{\alpha+2-\sigma} \leq C. \tag{H4}$$

A noncollision heteroclinic orbit at infinity is a solution of the problem

$$\begin{cases} \ddot{q} + V'(q) = 0 & \forall t \\ \lim_{t \to -\infty} q(t) = \lim_{t \to \pm\infty} \dot{q}(t) = 0 \\ \lim_{t \to +\infty} |q(t)| = 0 \\ q(t) \neq 0 & \forall t \end{cases} \quad (4.2)$$

Next theorem states sufficient conditions for the existence of such orbits.

THEOREM 4.2. ([21]). Assume V satisfies (H1)–(H4) with $N \geq 3$, $\beta \geq 2$ and $\alpha \in]0,2[$. Then problem (4.2) has at least one solution.

In this case the problem is to show the existence of minimizers to the functional naturally associated to (4.2), namely $f(q) = \frac{1}{2}\int_{-\infty}^{+\infty}|\dot{q}|^2 dt - \int_{-\infty}^{+\infty} V(q)dt$. The domain of f is a subset of the space

$$H = \{u \in H^1_{loc}(\mathbf{R}; \mathbf{R}^N) \,/\, \dot{u} \in L^2(\mathbf{R}; \mathbf{R}^N)\}.$$

Direct minimization of f does not yield a solution to (4.2). In [21], however it is shown that restricting f to

$$\Gamma_0^\infty = \{u \in H \,/\, \lim_{t \to -\infty} q(t) = 0, \lim_{t \to +\infty} |q(t)| = +\infty\}$$

there exists $q \in \Gamma_0^\infty$ minimizing f over this set. One then uses (H3) and (H4) to conclude, similarly to Theorem 3.1, that $q(t) \neq 0$ for all t.

C) THE FIXED ENERGY PROBLEM. We close this section with still another application of Theorem 3.1, concerning a fixed energy problem studied in [12]. The setting is the following.

Let $V \in C^2(\mathbf{R}^N \setminus \{0\}; \mathbf{R})$, let h be a fixed number and set $\Omega = \{x \in \mathbf{R}^N \,/\, V(x) < h\}$. We look for a special type of fixed energy solutions, namely the so-called "brake orbits" (see [7], [16]), i.e. solutions to

$$\begin{cases} \ddot{q} + V(q) = 0 \\ \dot{q}(0) = \dot{q}(T) = 0 \\ \frac{1}{2}|\dot{q}|^2 + V(q) = h \end{cases} \quad (4.3)$$

for all $t \in [0, T]$, T to be determined.

We assume that Ω has $k > 0$ holes, and precisely

$$\Omega = \Omega_0 \setminus (\cup_{i=1}^k \overline{\Omega}_i) \text{ is bounded and}$$
$$\overline{\Omega}_i \subset \Omega_0 \; \forall i, \quad \overline{\Omega}_i \cap \overline{\Omega}_j = \emptyset \; \forall i \neq j, \quad 0 \in \Omega, \quad (B1)$$

$$V'(x) \neq 0 \quad \forall x \in \partial\overline{\Omega} \tag{B2}$$

$$\exists r > 0, \quad \exists \phi \in C^2(]0, r[; \mathbf{R}) \quad \text{such that}$$
$$V(x) = -\frac{a}{|x|^\alpha} + \phi(|x|), \quad \forall 0 < |x| \leq r, \tag{B3}$$
$$\lim_{s \to 0} \phi'(s) s^{\alpha+1} = 0.$$

THEOREM 4.3. ([**12**]). *Let V satisfy (B1)–(B3), with $N \geq 3$, $a > 0$ and $\alpha \in]0, 2[$. Then there exist at least k noncollision solutions to problem (4.3).*

Remark that here we use (due to technical reasons) a slight variant, (B3), of (V2) and (V3). However the spirit of the proof is the same as above. We recall that solutions to (4.3) can be found as (rescalings of) critical points of the product functional

$$g(q) = \frac{1}{2} \int_0^1 |\dot{q}|^2 dt \cdot \int_0^1 [h - V(q)] dt$$

over $H^1([0, 1]; \mathbf{R}^N \setminus \{0\})$. Theorem 3.1 does not apply directly to g, but the result is true all the same. The key point is that, even though the functional setting is different, we still find solutions to (4.3) by *minimizing* g over classes of orbits linking different connected components of $\partial\Omega$. Once again, the minimizing property excludes the collisions.

For a complete discussion of the above results we refer to [**12**], [**21**], [**22**], and to [**23**] for further applications to the three–body problem.

REFERENCES

1. A. Ambrosetti and M. L. Bertotti, *Homoclinics for second order conservative systems*, Preprint SNS, 1991.
2. A. Ambrosetti and V. Coti Zelati, *Critical points with lack of compactness and singular dynamical systems*, Ann. Mat. Pura Appl. (4) **149** (1987), 237–259.
3. A. Ambrosetti and V. Coti Zelati, *Perturbations of Hamiltonian systems with Keplerian potentials*, Math. Zeit. **201** (1989), 227–242.
4. A. Ambrosetti and V. Coti Zelati, *Noncollision orbits for a class of Keplerian–like potentials*, Ann. IHP, Anal. non lin. **5** (1988), 287–295.
5. A. Ambrosetti and V. Coti Zelati, *Multiple homoclinic orbits for a class of conservative systems*, Preprint SNS, 1992.

6. A. Bahri and P.H. Rabinowitz, *A minimax method for a class of Hamiltonian systems with singular potential*, Jour. Funct. Anal. **82** (1989), 412–428.
7. V. Benci and F. Giannoni, *A new proof of the existence of a brake orbit*, Advanced topics in the theory of dynamical systems (Trento 1987), Notes Rep. Math. Sci. Energ. 6, Academic Press, 1990.
8. U. Bessi, *Multiple homoclinics orbits for autonomous, singular potentials*, Preprint SNS, 1992.
9. V. Coti Zelati, I. Ekeland and E. Séré, *A variational approach to homoclinic orbits in Hamiltonian systems*, Math. Ann. **288** (1990), 133–160.
10. V. Coti Zelati and P.H. Rabinowitz, *Homoclinic orbits for second order Hamiltonian systems possessing superquadratic potentials*, Jour. of AMS **4** (1991), 693–727.
11. V. Coti Zelati and E. Serra, *Collision and non-collision solutions for a class of Keplerian-like dynamical systems*, Ann. di Mat. Pura Appl. (to appear).
12. V. Coti Zelati and E. Serra, *Multiple brake orbits for some classes of Hamiltonian systems*, Nonlin. Anal. TMA (to appear).
13. M. Degiovanni and F. Giannoni, *Dynamical systems with Newtonian type potentials*, Ann. Sc. Norm. Sup. Pisa Cl. Sci. **4** (1989), 467–494.
14. W. Gordon, *Conservative dynamical systems involving Strong Forces*, Trans. AMS **204** (1975), 113–135.
15. C. Greco, *Periodic solutions of a class of singular Hamiltonian systems*, Nonlin. Anal. TMA **12** (1988), 259–269.
16. V.V. Kozlov, *Calculus of variations in the large and classical mechanics*, Russ. Math. Surv. **40** (1985), 37–71.
17. P.H. Rabinowitz, *Periodic and heteroclinic orbits for a periodic Hamiltonian system*, Ann. IHP **6** (1989), 331–346.
18. P.H. Rabinowitz and K. Tanaka, *Some results on connecting orbits for a class of Hamiltonian systems*, Math. Zeit. **206** (1991), 473–499.
19. E. Séré, *Existence of infinitely many homoclinic orbits in Hamiltonian systems*, Math. Zeit. **209** (1991), 27–42.
20. E. Séré, *Homoclinic orbits on compact hypersurfaces in $\mathbf{R}^{2N}$ of restricted contact type*, Preprint CEREMADE, 1992.
21. E. Serra, *Heteroclinic orbits at infinity for two classes of Hamiltonian systems*, Preprint, 1992.
22. E. Serra and S. Terracini, *Noncollision solutions to some singular minimization problems with Keplerian-like potentials*, Nonlin. Anal. TMA (to appear).
23. E. Serra and S. Terracini, *Noncollision periodic solutions to some three-body like problems*, Arch. Rat. Mech. Anal. (to appear).

24. K. Tanaka, *Non–collision solutions for a second order singular Hamiltonian system with weak forces*, Preprint, 1991.
25. K. Tanaka, *Homoclinic orbits for a singular second order Hamiltonian system*, Ann. IHP **7** (1990), 427–438.
26. S. Terracini, *An homotopical index and multiplicity of periodic solutions to dynamical systems with singular potentials*, J. Diff. Eq. (to appear).
27. S. Terracini, *Second order conservative systems with singular potentials: noncollision periodic solutions to the fixed energy problem*, Preprint.

PERIODIC SOLUTIONS OF HAMILTONIAN SYSTEMS

MICHAEL STRUWE

ETH Zürich, Mathematik, 8092 Zürich, Switzerland

1. Given a C^2-function $H\colon \mathbb{R}^{2n} \to \mathbb{R}$, we study periodic solutions $x\colon \mathbb{R} \to \mathbb{R}^{2n}$ of the Hamiltonian system

(1) $$\dot{x} = \mathcal{J}\nabla H(x),$$

where $\mathcal{J} = \begin{pmatrix} 0 & id \\ -id & 0 \end{pmatrix}$ is the standard symplectic structure on $\mathbb{R}^{2n} = \mathbb{R}^n \times \mathbb{R}^n$. By anti-symmetry of $\mathcal{J}$, we have $H(x(t)) =$ const. along a solution of (1). Thus, we may consider the question whether a given level surface $\{H(x) = \beta\}$ of the Hamiltonian H carries a periodic solution of (1).

Early results in this regard are due to Seifert [7] for Hamiltonians of a special form. Rabinowitz [5] and Weinstein [12] then answered this question affirmatively for level surfaces that are strictly convex or, more generally, strictly star-shaped. In 1987 Viterbo [11], moreover, showed that Rabinowitz' result remains true for regular, compact and connected level surfaces of "contact type", thus solving a conjecture of Weinstein [13] inspired by the work of Rabinowitz [5].

The notion of "contact-type" is natural from the point of view of dynamical systems and symplectic geometry. However, the question wheter periodic solutions to (1) exist on energy level-surfaces of more general type remained.

Extending Viterbo's results, Hofer-Zehnder [4] proved that in any neighborhood of a regular, compact and connected energy level surface there is a periodic solution of (1). Rabinowitz [6] observed that the proof of [4] may be adjusted to yield, in fact, uncountably many such solutions. The following result by this author [9], finally, shows that periodic solutions are even more abundant.

Theorem 1 *Suppose β_0 is a regular value of $H \in C^2\left(\mathbb{R}^{2n}\right)$ and that $H^{-1}(\beta_0)$ is compact and connected. Then for almost every β near β_0 there is a periodic solution of (1) on the energy level-surface $H^{-1}(\beta)$.*

The above results lends further support to the conjecture that a corresponding existence result holds true also at the level β_0 itself. Note, however, that such a result may be very sensitive to the geometry of $(\mathbb{R}^{2n}, \mathcal{J})$ and, in particular, does not hold for Hamiltonian systems on arbitrary symplectic manifolds; see Zehnder [14] ; p. 271f.

2. We sketch the proof of Theorem 1; see [9] for details. Basically, we use the approach of Hofer-Zehnder [4], slightly modified as in [10].

Note that wheter a level surface $S_\beta = H^{-1}(\beta)$ carries a periodic solution is a property of the surface, not of the particular Hamiltonian defined on it. Thus, for suitable $\delta > 0$, $|\alpha| < \delta$, $m \in \mathbb{N}$, and x such that $|H(x) - \beta_0| < 2\delta$ we define
$$\tilde{H}_{m,\alpha}(x) = f(m(H(x) - \beta_0 - \alpha)),$$
where $f(s) \equiv 0$ for $s \leq -\delta$, $f(s) \equiv b > 0$ for $s \geq \delta$, $f'(s) > 0$ for $|s| < \delta$, and we extend $\tilde{H}_{m,\alpha}$ suitably to $\mathbb{R}^{2n}$.

Then we consider the functional
$$E_{m,\alpha}(x) = \frac{1}{2}\int_0^1 \langle \dot{x}, \mathcal{J}x \rangle\, dt - \int_0^1 \tilde{H}_{m,\alpha}(x)\, dt,$$
defined on the Sobolev space $V = H^{\frac{1}{2},2}(S^1; \mathbb{R}^{2n})$ of closed loops
$$x = \sum_{k \in \mathbb{Z}} x_k \exp(2\pi \mathcal{J} k t)$$
with Fourier coefficients $x_k \in \mathbb{R}^{2n}$ satisfying $\|x\|^2 := |x_0|^2 + \Sigma_{k \in \mathbb{Z}}|k|\|x_k\|^2 < \infty$. One verifies that $E_{m,\alpha} \in C^2\left(H^{\frac{1}{2},2}(S^1; \mathbb{R}^{2n})\right)$ and that critical points x of $E_{m,\alpha}$ with $E_{m,\alpha}(x) > 0$ may only occur in the energy region $|H(x(t)) - \beta_0| < 2\delta$. Thus, these x will satisfy the equation
$$\dot{x} = \mathcal{J}\nabla \tilde{H}_{m,\alpha}(x) = T\mathcal{J}\nabla H(x),$$
where
$$T = T(x) = mf'(m(H(x) - \beta_0 - \alpha)).$$
That is, x gives rise to a T-periodic solution $y(t) = x\left(\frac{t}{T}\right)$ of (1). Observe that, on the other hand
$$T = \frac{\partial}{\partial \alpha} E_{m,\alpha}(x),$$
and thus the periods $T(x_m)$ of a sequence of solutions x_m for $E_{m,\alpha}$, say, may be controlled if $\frac{\partial}{\partial \alpha}E_{m,\alpha}(x_m)$ can be controlled. The latter control is achieved using a variational technique, developed by the author, for instance, in [8];

p. 60, (4.8)–(4.10), based on monotonicity of $E_{m,\alpha}$ in α. The same technique in a related context was recently used by Ambrosetti-Bertotti [1].

As in Hofer-Zehnder [4] the space $H^{\frac{1}{2},2}\left(S^1; \mathbb{R}^{2n}\right) = V$ splits into sub-spaces $V = V^+ \oplus V^0 \oplus V^-$ which are invariant for the gradient of the action

$$A(x) = \frac{1}{2} \int_0^1 \langle \dot{x}, \mathcal{J}x \rangle \, dt.$$

Since the gradient of the Hamiltonian integral

$$G_{m,\alpha}(x) = \int_0^1 \tilde{H}_{m,\alpha}(x(t)) \, dt$$

is compact, the gradient flow of $E_{m,\alpha}$ preserves the sub-spaces $V^+ \oplus V^0 \oplus V^-$ up to a compact error term.

Define Γ to be the class of maps $h \in C^0(V; V)$ homotopic to the identity through a family of maps $h_t = L_t + K_t$, $0 \leq t \leq T$, where $L_0 = \mathrm{id}$, $K_0 = 0$, and where for each $t \in [0, T]$ the map K_t is compact while

$$L_t = \left(L_t^+, L_t^0, L_t^-\right) : V^+ \oplus V^0 \oplus V^- \to V^+ \oplus V^0 \oplus V^-$$

is a linear isomorphism preserving the sub-spaces V^+, V^0, V^-. Note that, for instance, the gradient flow $\Phi_{m,\alpha}$ of $E_{m,\alpha}$ for any $|\alpha| < \delta$, $m \in \mathbb{N}$ induces a family of maps $\Phi_{m,\alpha}(\cdot, t) \in \Gamma$.

Let

$$S_\varrho^+ = \{x \in V^+; \|x\| = \varrho\},$$

and for fixed $e = \frac{1}{\sqrt{2\pi}} \exp(2\pi \mathcal{J} t) a \in V^+$, $|a| = 1$, and $R > 0$ define

$$Q_R = \{x = x^- + x^0 + se \in V^- \oplus V^0 + \mathbb{R}e; \; \|x^- + x^0\| \leq R, \; 0 \leq s \leq R\}$$

with relative boundary

$$\partial Q_R = \left\{x \in Q; \; \|x^- + x^0\| = R \;\; \text{or} \; s \in \{0, R\}\right\}.$$

For suitably large $R > 0$ there holds

$$E_{m,\alpha} \leq 0 \quad \text{on} \quad \partial Q_R,$$

for all $m \in \mathbb{N}$, $|\alpha| < \delta$. On the other hand for small $\varrho > 0$ there exists $\mu_0 > 0$ such that

$$E_{m,\alpha}(x) \geq \mu_0 > 0 \quad \text{for all } x \in S_\varrho^+,$$

independently of m and α. Fix such numbers $0 < \varrho < R$, $\mu_0 > 0$ and let $Q = Q_R$ accordingly. Let $\Gamma_{m,\alpha} \subset \Gamma$ be the sub-class

$$\Gamma_{m,\alpha} = \left\{ h \in \Gamma;\ E_{m,\alpha}\big|_{h_t(\partial Q)} \leq 0\ \text{ for all } t \in [0,T] \right\},$$

where h_t is a homotopy as in the definition of Γ. Then ∂Q and S_ϱ^+ link with respect to $\Gamma_{m,\alpha}$ in the sense of Benci-Rabinowitz [3], and the associated numbers

$$\mu_m(\alpha) = \inf_{h \in \Gamma_{m,\alpha}} \sup_{x \in Q} E_{m,\alpha}(h(x)) \geq \mu_0$$

are critical values of $E_{m,\alpha}$ for any $m \in \mathbb{N}$, $|\alpha| < \delta$. The choice $h = id$ as comparison function also yields a uniform upper bound $\mu_{m,\alpha} \leq \mu_\infty$ for all m and α.

Moreover, since $E_{m,\alpha'} \leq E_{m,\alpha}$ and thus also $\Gamma_{m,\alpha} \subset \Gamma_{m,\alpha'}$ for $\alpha' \leq \alpha$, we have

$$\mu_m(\alpha') \leq \mu_m(\alpha)\ \text{ for }\ \alpha' \leq \alpha.$$

In particular, the functions $\alpha \mapsto \mu_m(\alpha)$ are almost everywhere differentiable, for any $m \in \mathbb{N}$, and we have the uniform BV-estimate

$$\int_{-\delta}^{\delta} \left|\frac{\partial}{\partial \alpha}\mu_m(\alpha)\right| d\alpha = \int_{-\delta}^{\delta} \frac{\partial}{\partial \alpha}\mu_m(\alpha)\, d\alpha \leq \mu_\infty - \mu_0.$$

By Fatou's lemma, finally, this implies

$$\liminf_{m \to \infty} \frac{\partial}{\partial \alpha}\mu_m(\alpha) \in L^1([-\delta, \delta])$$

with

$$\int_{-\delta}^{\delta} \liminf_{m \to \infty} \frac{\partial}{\partial \alpha}\mu_m(\alpha)\, d\alpha \leq \mu_\infty - \mu_0.$$

The key lemma in the proof of Theorem 1 now is the following.

Lemma 1 *Suppose $|\alpha_0| < \delta$ is such that $C_0 := \liminf_{m \to \infty} \frac{\partial}{\partial \alpha}\mu_m(\alpha_0) < \infty$. Then there is a sequence $\Lambda \subset \mathbb{N}$ and corresponding critical points x_m of E_{m,α_0} such that*

$$E_{m,\alpha_0}(x_m) = \mu_m(\alpha_0)\ \text{ and }\ T(x_m) \leq C_0 + 4,\ m \in \Lambda.$$

Lemma 1 is based on the observation that for $\alpha_k > \alpha_0$, $|\alpha_k - \alpha_0| \ll 1$, the gradient flow for E_{m,α_0} also is a pseudo-gradient flow for E_{m,α_k} on the set where

$$\mu_m(\alpha_0) - 2(\alpha_k - \alpha_0) \leq E_{m,\alpha_0}(x) \leq E_{m,\alpha_k}(x)$$
$$\leq \mu_m(\alpha_k) + (\alpha_k - \alpha_0) \leq \mu_m(\alpha_0) + (C_0 + 2)(\alpha_k - \alpha_0).$$

To complete the proof of Theorem 1 remark that the sequence (x_m) of Lemma 1 satisfies the conditions

$$\dot{x} = T(x_m)\mathcal{J}\nabla H(x_m),$$
$$|H(x_m) - \beta_0 - \alpha_0| \to 0$$
$$A(x_m) \geq E_{m,\alpha_0}(x_m) \geq \mu_0 > 0.$$

Since $T(x_m) \leq C_0 + 4$, by the theorem of Arzéla-Ascoli we may assume that (x_m) converges C^1-uniformly to a 1-periodic solution x of

$$\dot{x} = T\mathcal{J}\nabla H(x)$$

with $H(x) = \beta_0 + \alpha_0$, $A(x) \geq \mu_0$. In particular, $x \neq \text{const.}$ The proof is complete, as Lemma 1 applies to almost every $\alpha \in \,]-\delta, \delta[$.

References

[1] Ambrosetti, A., Bertotti, M.L.: Homoclinics for second order conservative systems. Scuola Norm. Sup. Pisa, Preprint 107, June 1991

[2] Benci, V., Hofer, H., Rabinowitz, P.H.: A priori bounds for periodic solutions on hypersurfaces, in: Periodic solutions of Hamiltonian systems and related topics (eds. Rabinowitz et al.), Reidel 1987

[3] Benci, V., Rabinowitz, P.H.: Critical points theorem for indefinite functionals. Invent. Math. **52** (1979) 241–273

[4] Hofer, H., Zehnder, E.: Periodic solutions on hypersurfaces and a result by C. Viterbo. Invent. Math. **90** (1987) 1–9

[5] Rabinowitz, P.H.: Periodic solutions of Hamiltonian systems. Comm. Pure Appl. Math. **31** (1978) 157–184

[6] Rabinowitz, P.H.: On a theorem of Hofer and Zehnder, in: Periodic solutions of Hamiltonian systems and related topics, NATO ASI, Ser. C, vol. 209 (1987) 245–255

[7] Seifert, H.: Periodische Bewegungen mechanischer Systeme. Math. Z. **51** 197–216

[8] Struwe, M.: The existence of surfaces of constant mean curvature with free boundaries. Acta Math. **160** (1988) 19–64

[9] Struwe, M.: Existence of periodic solutions of Hamiltonian systems on almost every energy surface. Bol. Soc. Bras. Mat. **20** (1990) 49–58

[10] Struwe, M.: Variational methods. Springer, Berlin Heidelberg (1990)

[11] Viterbo, C.: A proof of the Weinstein conjecture in $\mathbb{R}^{2n}$. Ann. Inst. H. Poincaré, Analyse Non Linéaire **4** (1987) 337–356

[12] Weinstein, A.: Periodic orbits for convex Hamiltonian systems. Ann. Math. **108** (1978) 507–518

[13] Weinstein, A.: On the hypotheses of Rabinowitz' periodic orbit theorem. J. Diff. Eq. **33** (1979) 353–358

[14] Zehnder, E.: Remarks on periodic solutions on hypersurfaces in: Periodic solutions of Hamiltonian systems and related topics (ed. P. H. Rabinowitz), D. Reidel (1987) 267–279

A NOTE ON PALAIS-SMALE CONDITION AND MOUNTAIN-PASS PRINCIPLE FOR LOCALLY LIPSCHITZ FUNCTIONALS

STEPAN TERSIAN

Technical University, Centre of Mathematics, Rousse 7017, Bulgaria

Abstract. Some aspects of critical point theory for locally Lipschitz functionals over a Banach space are discussed in the work. The minimization, coercivity and mountain pass principle are considered using Ekeland's variational principle.

INTRODUCTION

In this paper we consider some aspects of minimization of locally Lipschitz functionals on Banach spaces. It is well known that generalized gradients can be defined (cf. Clarke[5]) for such functionals. A general approach by variational methods for locally Lippschitz functionals was developed by Chang[4] with the use of the generalized gradient of Clarke.

In this note we reformulate at first the well known Ekeland's variational principle for such functionals in terms of directional derivatives. In a strong form it can be stated as follows:

Theorem 1 .*(Ekeland) Let (M, d) be a complete metric space and $\Phi : M \to \mathbf{R}$ be a lower semicontinuous function which is bounded from bellow and $x_0 \in M$, $\epsilon > 0$ are given. Then there exists $y_0 \in M$ such that:*

　　i)　　$\Phi(y_0) + \epsilon d(x_0, y_0) \leq \Phi(x_0),$
　　ii)　　$\Phi(y_0) < \Phi(x) + \epsilon d(x, y_0), \forall x \in M, x \neq x_0.$

Next we introduce a variant of Palais-Smale (*PS*) condition which we prove to be equivalent to the condition introduced by Chang[4]. This allows to formulate general minimization theorem for locally Lipschitz functionals as well as some results for coercitivity and mountain pass principle for such functionals.

ABSTRACT FRAMEWORK: MAIN RESULTS

Let X be a Banach space, X^* is its dual space, $\|.\|$ is the norm and $<p^*, x>$ for $p^* \in X^*, x \in X$ denotes the duality. Let $f : X \to \mathbf{R}$ be a locally Lipschitz functional, i.e. for each $x \in X$ there exists a neigbourhood $N(x)$ of x and a consant L depending on N such that $\|f(y) - f(z)\| \leq L\|y - z\|, \forall y, \forall z \in N(x)$.

We'll use notation $f \in LL(X)$ for such functionals.

For each $v \in X$ it can be defined directional derivative

$$D_c f(x)(v) = \limsup_{h \to 0_+, y \to x} \frac{f(y + hv) - f(y)}{h}$$

$$= \inf_{\delta > 0} \sup_{\|y-x\|<\delta, 0<h<\delta} \frac{f(y + hv) - f(y)}{h}.$$

We have the following basic properties of $D_c f(x)(v)$- Aubin [2]:

1) The function $v \to D_c f(x)(v)$ is subaddiative, positively homogeneous and then convex.
2) $|D_c f(x)(v)| \leq L\|v\|, \forall v \in X$,
3) The function $v \to D_c f(x)(v)$ is upper semicontinuous, i.e.

$$\limsup_{x \to x_0, v \to v_0} D_c f(x)(v) \leq D_c f(x_0)(v_0)$$

4) The function

$$\lambda(x) = \min\{\|p\|_* : <p, v> \leq D_c f(x)(v)\}$$

exists, and is lower-semicontinuous, i.e.

$$\liminf_{x \to x_0} \lambda(x) \geq \lambda(x_0).$$

5) There exists $p_0 \in X^* : D_c f(x)(v) = <p_0, v>, \forall v \in X$.

Recall the definition of the generalized gradient due to F. Clarke of f at x, denoted by $\partial f(x)$: $p \in \partial f(x)$ iff $<p, v> \leq D_c f(x)(v), \forall v \in X$.

We have

$$\lambda(x) = \min\{\|p\|_* : p \in \partial f(x)\}.$$

The reformulation of Ekeland's principle for locally Lipschitz functionals is as follows.

Theorem 2 . Let $f : X \to \mathbf{R} \cup \{\infty\}$ be locally Lipschitz function which is bounded from bellow and $x_0 \in \text{Dom} f$. Then there exists $y_0 \in X$:

$$\text{i*}) \qquad f(y_0) + \epsilon\|x_0 - y_0\| \leq f(x_0),$$
$$\text{ii*}) \qquad 0 \leq D_c f(x)(v) + \epsilon\|v\|, \forall v \in X.$$

Proof. We have by condition $i)$ of Theorem 1:

$$0 < f(x) - f(y_0) + \epsilon\|x - y_0\|$$

and for $x = y_0 + hv, h > 0, v \in X$:

$$0 < \frac{f(y_0 + hv) - f(y_0)}{h} + \epsilon\|v\|$$

Let $\epsilon_1 > 0$ be an arbitrary positive and $\delta_1 > 0$:

$$\sup_{\|y-y_0\|<\delta_1, 0<h<\delta_1} \frac{f(y+hv) - f(y)}{h} \leq D_c f(y_0)(v) + \epsilon_1.$$

Then for $h < \delta_1, \forall v \in X$:

$$0 \leq \frac{f(y_0 + hv) - f(y_0)}{h} + \epsilon\|v\|$$
$$\leq \sup_{\|y-y_0\|<\delta_1, 0<h<\delta_1} \frac{f(y+hv) - f(y)}{h} + \epsilon\|v\|$$
$$\leq D_c f(x)(v)) + \epsilon\|v\| + \epsilon_1.$$

As $\epsilon_1 > 0$ is arbitrary:

$$0 \leq D_c f(x)(v)) + \epsilon\|v\|. \quad \Box$$

Next recall that x_0 is a critical point of f if

$$0 \leq D_c f(x)(v), \forall v \in X$$

or $0 \in \partial f(x_0)$. In Chang [4] introduced the following Palais- Smale (PS) condition for locally Lipschitz functionals, which reduces to the usual Palais-Smale condition when f is C^1 functional:

Definition 1 . The functional $f : X \longrightarrow R$ satisfies (PS^0) condition if whenever $\{x_n\} \subset X$ is such that:

$$\text{j}) \qquad |f(x_n)| \text{ is bounded,}$$
$$\text{jj}) \qquad \lambda(x_n) = \min\{\|p_n\|_* : p_n \in \partial f(x_n)\} \longrightarrow 0,$$

then $\{x_n\}$ posseses a convergent subsequence .

We formulate the following definition for (PS^1) condition.

Definition 2. *The functional $f \in LL(X)$ is said to satisfy (PS^1) condition if when ever $\{x_n\} \subset X$ is such that:*

$$j^*) \quad |f(x_n)| \text{ is bounded,}$$
$$jj^*) \quad \forall \epsilon > 0, \exists n_0 : n > n_0 \Rightarrow$$
$$\leq D_c f(x)(v) + \epsilon \|v\|, \forall v \in X$$

then $\{x_n\}$ posseses a convergent subsequence.

Lemma 1. *The definitions of (PS^0) and (PS^1) conditions are equivalent.*

Proof. $\Rightarrow$ part. Let

$$\lambda(x_n) = \min\{p_n \in \partial f(x_n)\} = \|p_n\|_* \longrightarrow 0, \text{ and } \epsilon > 0.$$

Then $\exists n_0 : n > n_0 \Rightarrow \|p_n\|_* < \epsilon \Leftrightarrow |<p_n,v>| \leq \epsilon\|v\| \forall v \in X$. We have $-|<p_n,v>| \leq |<p_n,v>| \leq \epsilon\|v\|$ or $0 \leq <p_n,v> + \epsilon\|v\| \leq D_c f(x_n) + \epsilon\|v\|$ and $(jj*)$ is satisfied.

$\Leftarrow$ part. Let now $\epsilon > 0$ and n_0 is such that for $n > n_0$: $0 \leq D_c f(x_n) + \epsilon\|v\|$ and $P_n \in \partial f(x_n)\}$ is such that $<p_n,v> = D_c f(x_n)$. Then $0 \leq <p_n,v> + \epsilon\|v\|, \forall v \in X$. For $-v \in X \Rightarrow 0 \leq - <p_n,v> + \epsilon\|v\|$ and then $|<p_n,v>| \leq \epsilon\|v\|$ or

$$\lambda(x_n) \leq \|p_n\|_* \text{ for } n > n_0.$$

that is (jj) is satisfied. $\square$

Note that (jj^*) in (PS^1) can be replaced by
(jj^*) There exists $\epsilon_n \longrightarrow 0$ as $n \longrightarrow \infty$ such that

$$0 \leq D_c f(x_n)(v) + \epsilon_n \|v\|, \forall v \in X.$$

Now we can prove

Theorem 3. *Let $f : X \to \mathbf{R} \cup \{\infty\}$ be a locally Lipschitz functional which is bounded from bellow and satisfies (PS^1) condition. Then there exists x_0 such that $f(x_0) = \inf f(x)$ and x_0 is a critical point i.e.: $0 \leq D_c f(x_0)(v) + \epsilon_n \|v\|, \forall v \in X$.*

Proof. Let $\{x_n\}$ be a minimizing sequence, i.e.

$$f\{x_n\} \longrightarrow \inf f \text{ as } n \to \infty.$$

Let $\epsilon > 0$ be an arbitrary number and for $n > n_0$

$$\inf f \leq f(x_n) \leq \inf f + \epsilon.$$

By Theorem 2 there exists y_n :

$$\inf f \leq f(y_n) \leq f(x_n) + \epsilon.$$
$$0 \leq D_c f(y_n)(v) + \epsilon\|v\|, \forall v \in X.$$

By (PS) condition there exists a convergent subsequence $\{y_{n_k}\}$ of $\{y_n\}$ and let $\{y_{n_k}\} \to x_0$. By upper semicontinuity of $D_c f(.)(.)$ it follows

$$0 \leq D_c f(x_0)(v) + \epsilon\|v\|, \forall v \in X.$$

and $\inf f \leq f(x_0) \leq \inf f + \epsilon$. As ϵ is arbitrary: $f(x_0) = \inf f$ and $0 \leq D_c f(x_0)(v) \forall v \in X$, i.e. x_0 is a critical point and of global minimum. □

Now let's reformulate Ekeland's principle as

Theorem 4 . *Let $f \in LL(X)$ be a function which is bounded from bellow and $x_0 \in X$ be given such that $f(x_0) \leq \inf f + \epsilon$. Then for every $\lambda > 0$ there exists $y_0 \in X$:*

1) $f(y_0) \leq f(x_0)$
2) $\|x_0 - y_0\| \leq \dfrac{1}{\lambda}$
3) $0 \leq D_c f(y_0)(v) + \lambda\epsilon\|v\|, \forall v \in X$.

The condition follows by replacing $\|.\|$ with the equivalent norm $\lambda\|.\|$. If $\lambda = 1/\sqrt{\epsilon}$ we have

1) $f(y_0) \leq f(x_0)$
2) $\|x_0 - y_0\| \leq \sqrt{\epsilon}$
3) $0 \leq D_c f(y_0)(v) + \sqrt{\epsilon}\|v\|, \forall v \in X$.

Now we shall discuss the coercivity of the bounded from below locally Lipschitz functonals. Recall that $f : X \longrightarrow \mathbf{R}$ is coercive if $f(x) \to +\infty$ as $\|x\| \to \infty$. We prove the followng.

Theorem 5 . *Let X be a Banach space and $f \in LL(X)$ be a functional satisfying (PS^1) condition. If f is bounded from bellow, then f is coercive.*

This result is a generalizaton of those proved by Caklovich, Li and Willem[4]. We follow their method of proof.

Proof. Suppose that the conclusion is not true and $c = \liminf_{\|x\|\to\infty} f(x)$ is finite. Let for $\epsilon = 1/n$ there exists $x_n : \|x_n\| \geq 2n$

$$f(x_n) \leq c + \epsilon = \inf f + (c + \frac{1}{n} - \inf f).$$

By Theorem 2 there exists $y_n \in X$:

$$f(y_n) \leq f(x_n),$$
$$\|x_n - y_n\| \leq n,$$
$$0 \leq D_c f(y_n)(v) + \frac{1}{n}(c + \frac{1}{n} - \inf f)\|v\|, \forall v \in X.$$

We have $\|y_n\| \geq \|x_n\| \geq 2n - n = n$, i.e. $\lim_{n \to \infty} \|y_n\| = \infty$. On the other hand by

$$\epsilon_n = \frac{1}{n}(c + \frac{1}{n} - \inf f) \to 0 \text{ as } n \to \infty$$

and (PS') condition there exists a convergent subsequence of $\{y_n\}$, which is a contradiction. Then $c = +\infty$. □

Let us denote by f_c the set $\{x \in X : f(x) \leq c\}$. In the sequel we prove

Theorem 6 . *Let $f : X \to \mathbf{R}$ be a locally Lipschitz function which is bounded from bellow and satisfies (PS^1) condition. If $c = \inf f$ there exists $a > 0$ such that f_{c+a} is a bounded set.*

Proof. Assume the contrary:

$$\forall \alpha > 0, \forall R > 0 \exists x : f(x) \leq c + \alpha \text{ and } \|x\| \geq R.$$

Let for $\alpha = \frac{1}{n^2}, R = 2n, \exists x_n : \|x_0\| \geq 2n$ and $f(x_0 \leq c + \frac{1}{n^2}$. There exists y_n by Theorem 4 such that:

$$f(y_n) \leq f(x_n)$$
$$\|x_n - y_n\| \leq 1/n$$
$$0 \leq D_c f(y_n)(v) + \frac{1}{n}\|v\|, \forall v \in X.$$

Then $\|y_n\| \geq \|(x_n)\| - \|(x_n) - y_n\| \geq 2n - 1/n > n$ and $\lim_{n \to \infty} \|y_n\| = +\infty$. On the other hand by (PS) condition $\{y_n\}$ is precompact and we get a contradiction. Then there exists $a > 0 : f_{c+a}$ is bounded. □

Note that we can formulate a weaker variant of (PS) condition known as $(PS)_c$ condition as follows:

Definition 3 . *The function $f \in LL(X)$ satisfies $(PS)_c$ condition if for $c \in bfR$ and any sequence $\{x_n\}$ such that*

j$_c$) $f(x_n) \to c$ as $n \to \infty$,
jj$_c$) $\forall \epsilon > 0, \exists n_0 : n > n_0 \Rightarrow 0 \leq D_c f(x_n)(v) + \epsilon\|v\|, \forall v \in X,$

it follows that $\{x_n\}$ is precompact.

Note that in Theorem 3 and 5 (PS^1) condition can be replaced by a weaker one - $(PS^1)_c$ with $c = \inf f$. Our next result in this direction is as follows:

Theorem 7. *Let Let $f \in LL(X)$ be a function bounded from bellow and $b > \inf f$ is such that f_b is a bounded set. If f satisfies $(PS^1)_b$ condirion then there exists $a > 0$ such that f_{b+a} is also bounded.*

Proof. Assume the contrary: $\forall \alpha > 0, \forall R > 0, \exists x \in f_{b+\epsilon}$ and $\|x\| \geq R$. Let x_n is such that

$$f(x_n) \leq b + 1/n \text{ and } \|x_n\| \geq 2n.$$

It follows that $\|y_n\| \geq \|x_n\| - \|x_n - y_n\| \geq n$ and $\lim \|y_n\| = +\infty$. There exists $n_1 : n > n^1 \Rightarrow f(y_n) > b$; otherwise $y_n \in f_b$ which is bounded - a contradiction! Then $f(y_n) \to b$ as $n \to +\infty$ and by $(PS^1)_b$ condition $\{y_n\}$ admits a convergent subsequence which contradicts to $\|y_n\| = +\infty$. Then there exists $a : f_{b+a}$ is bounded. □

MOUNTAIN PASS PRINIPLE FOR LOCALLY LIPSCHITZ FUNCTIONALS

Let's consider the space of paths joinning two points 0 and e of X:

$$\Gamma = \Gamma(0, e) = \{c \in C([0,1], X) : c(0) = 0, c(1) = e\}$$

equiped with the uniform norm $[[.]]$:

$$[[c_1 - c_2]] = \max\{\|c_1(t) - c_2(t)\| : 0 \leq t \leq 1\}.$$

Let $f \in LL(X)$ be a functional and define the new functional over Γ

$$I(c) = \max\{|f(c(t))| : t \in [0,1]\}$$

over the space Γ. We have

Lemma 2. *If $f \in LL(X)$, then $I \in LL(\Gamma)$.*

Proof. Let $c \in \Gamma$. We prove first that

$$\exists \delta > 0, \exists L > 0 : \forall x \in c, \forall y, z \in x + \frac{\delta}{2}B \Rightarrow |f(y) - f(z)| \leq L\|y - z\|. \quad (1)$$

Since $f \in LL(X)$

$$\forall x \in c, \exists \delta_x > 0, \exists L_x > 0 : \forall y, z \in x + \delta_x B \Rightarrow |f(y) - f(z)| \leq L_x \|y - z\|.$$

Note that B is the unit ball in X. Since c is a compact in X and $c \subset \bigcup \{x + \frac{\delta_x}{2} B : x \in c\}$ there exist $\{x_i, \delta_i, L_i : i = 1, \ldots, n\}$ such that $c \subset \bigcup \{x_i + \frac{\delta_i}{2} B : i = 1, \ldots, n\}$ and $|f(y) - f(z)| \leq L_i \|y - z\|$, whenever $\forall y, z \in x_i + \delta_i B$. Let now $\delta = \min\{\delta_i : i = 1, \ldots, n\}$, $L = \max\{L_i : i = 1, \ldots, n\}$ and $x \in c, \forall y, z \in x + \frac{\delta}{2} B$. There exists $x_j : x \in x_j + \frac{\delta_j}{2} B :$ and then

$$\|y - x_j\| \leq \|y - x\| + \|x - x_j\| \leq \frac{\delta}{2} + \frac{\delta_j}{2} \leq \delta_j,$$
$$\|z - x_j\| \leq \delta_j \Rightarrow |f(y) - f(z)| \leq L_j \|y - z\| \leq L \|y - z\|.$$

Now we prove that

$$\forall c_1, c_2 \in \Gamma : [[c_k - c]] < \delta/2, k = 1, 2 \Rightarrow |I(c_1) - I(c_2)| \leq [[c_1 - c_2]] \quad (2)$$

Note that

$$\begin{aligned}|I(c_1) - I(c_2)| &= |\max |f(c_1(t))| - \max |f(c_2(t))|| \\ &\leq \max |f(c_1(t)) - f(c_2(t))| \\ &= |f(c_1(t_1)) - f(c_2(t_1))| = A_1.\end{aligned}$$

Then by $\|c_k(t_1)) - c(t_1)\| \leq [[c_k - c]] < \delta/2, k = 1, 2$ and (1):

$$A_1 \leq L \|c_1(t_1)) - c_2(t_1)\| \leq L[[c_1 - c_2]]$$

$$|I(c_1) - I(c_2)| \leq [[c_1 - c_2]]. \quad \square$$

Now we will discuss a generalization of montain-pass theorem of Ambrosetti-Rabinowitz for locally Lipschitz functions. We are following the ideas developed in Aubin, Ekeland [2]. First let us formulate another weak form of (PS) condition introduced in Aubin, Ekeland [2].

Definition 4 . *The function $f \in LL(X)$ satisfies (WPS) condition if whenever $\{x_n\}$ is a sequence such that*

j$_w$) $f(x_n)$ *is bounded*
jj$_w$) $\lambda(x_n) = \min\{\|p\| : p \in \partial f(x_n) \to 0\}, \lambda(x_n) \neq 0$,

it follows that there exists x_0 :

$$\liminf f(x_n) \leq f(x_n) \leq \limsup f(x_n).$$

Now we have

Theorem 8 Let $f \in LL(X)$ be a function and $\exists \rho > 0$ and $e \in X$:

1) $m(\rho) = \inf\{|f(x)| : \|x\| = \rho\} > f(0)$,
2) $\|e\| > \rho, f(e) < m(\rho)$,
3) f satisfies (WPS) condition.

Then there exists x_0:
$$f(x_0) \geq m(\rho), \lambda(x_0) = 0.$$

Proof. Lets's consider the functional I over the space Γ. $I \in LL(\Gamma)$ and is bounded from bellow by $m(\rho)$. Indeed if $c \in \Gamma, \exists t_0 : \|c(t_0)\| = \rho$ and then
$$I(c) \geq |f(c(t_0))| \geq m(\rho).$$
By Ekeland's theorem for $\epsilon > 0, \exists c_\epsilon \in \Gamma$:

$$I(c_\epsilon) \leq \inf I(c) + \epsilon, \tag{3}$$
$$0 \leq I(c) - I(c_\epsilon) + \epsilon[[c_\epsilon - c]], \forall c \in \Gamma. \tag{4}$$

Let $\gamma \in C([0,1], X) : \gamma(0) = \gamma(1) = 0, [[\gamma]] \leq 1$. Then if $h > 0, c \in \Gamma, c_h = c + h\gamma \in \Gamma$ and $[[c_h - c]] \leq h$ there exists h_0 and $L > 0 : h < h_0$ such that

$$|I(c_{\epsilon,h}) - I(c_\epsilon)| \leq [[c_{\epsilon,h} - c_\epsilon]] \leq Lh \leq Lh_0$$
$$I(c_{\epsilon,h}) \leq I(c_\epsilon) + |I(c_{\epsilon,h}) - I(c_\epsilon)| \tag{5}$$
$$\leq \inf I + \epsilon + Lh_0;$$
$$\exists h_0 > 0, L > 0 : \forall h < h_0, c_{\epsilon,h} = c_\epsilon + h\gamma \Longrightarrow$$
$$I(c_{\epsilon,h}) \leq \inf I + \epsilon + Lh_0. \tag{6}$$

We have by (4) for $c_{\epsilon,h} = c_\epsilon + h\gamma$ with $h < h_0$:

$$0 \leq 1/h(I(c_\epsilon + h\gamma) - I(c_\epsilon)) + \epsilon[[\gamma]]. \tag{7}$$

Let us define the function $\Phi : C[0,1] \longrightarrow \mathbf{R}$

$$\Phi(\phi) = \max_{t \in [0,1]} |\phi(t)|.$$

Φ is a convex and continuous functional and then admits a subdifferential $\partial \Phi(.) : \mu \in \partial \Phi(\phi_0) \subset C[0,1]^*$

$$\Phi(\phi) - \Phi(\phi_0) \geq <\mu, \phi - \phi_0>_C = \int(\phi - \phi_0)d\mu,$$

where the conjugate space $C[0,1]^*$ is the space of Radon measures μ over $[0,1]$ and
$$\partial\Phi(\phi) = \{\mu \geq 0 : \int d\mu = 1, supp\mu \subset M(\phi)\},$$
where
$$M(\phi) = \{t : \phi(t) = \Phi(\phi).$$
By Leburg theorem - Clarke[5], Th.2.3.7 we have
$$f(c_\epsilon(t) + h\gamma(t)) - f(c_\epsilon(t)) = h <p_\epsilon, \gamma(t)>,$$
where $p_\epsilon \in \partial f(c_\epsilon(t) + h_\epsilon \gamma(t))$ with $0 < h_\epsilon < h < h_0$. We have by (7):
$$\begin{aligned}
0 &\leq 1/h(I(c_\epsilon + h\gamma) - I(c_\epsilon)) + \epsilon[[\gamma]] \\
&= 1/h(\max_t |f(c_\epsilon(t) + h\gamma(t))| - \max_t |f(c_\epsilon(t))|) + \epsilon[[\gamma]] \\
&= 1/h(\max_t |f(c_\epsilon(t) + h <p_\epsilon, \gamma(t)>| - \max_t |f(c_\epsilon(t))|) + \epsilon[[\gamma]] \\
&= 1/h(\Phi(f(c_\epsilon + h <p_\epsilon, \gamma>) - \Phi(f(c_\epsilon))) + \epsilon[[\gamma]] \\
&= max_\mu \{\int <p_\epsilon, \gamma> d\mu : \mu \in \partial\Phi(f(c_\epsilon))\} + \epsilon[[\gamma]].
\end{aligned}$$

and then
$$-\epsilon \leq \inf_\gamma \max_\mu \left\{ \begin{array}{c} \mu \in \partial\Phi(f(c_\epsilon)) \\ \int <p_\epsilon, \gamma> d\mu : [[\gamma]] \leq 1 \\ \gamma(0) = \gamma(1) = 0 \end{array} \right\} = A_2.$$

As $\partial\Phi(f(c_\epsilon))$ is a weak* compact by minimax theorem Aubin, Ekeland[2], Th.6.2.7:
$$\begin{aligned}
-\epsilon \leq &= A_2 = \max_\mu \inf_\gamma \left\{ \begin{array}{c} \mu \in \partial\Phi(f(c_\epsilon)) \\ \int <p_\epsilon, \gamma> d\mu : [[\gamma]] \leq 1 \\ \gamma(0) = \gamma(1) = 0 \end{array} \right\} \\
&= \max_\mu \left\{ -\int \|p_\epsilon\|_* d\mu : \mu \in \partial\Phi(f(c_\epsilon)) \right\} \\
&= -\min\{\|p_\epsilon(c_\epsilon(t) + h_\epsilon\gamma(t))\|_* : t \in M(f(c_\epsilon(t) + h_\epsilon\gamma(t)))\}
\end{aligned}$$

There exists t_ϵ such that for $x_\epsilon = c_\epsilon(t_\epsilon) + h_\epsilon\gamma(t_\epsilon)$:
$$|f(x_\epsilon)| = \max_t |f(c_\epsilon(t) + h_\epsilon\gamma(t))|, \tag{8}$$
$$\|\lambda(x_\epsilon)\| \leq \|p_\epsilon(x_\epsilon)\|_* \leq \epsilon. \tag{9}$$

Let now $\epsilon = 1/n$ and $u_n = x_{1/n}$. Then by (9)
$$\|\lambda(u_n)\|_* \leq 1/n.$$

Moreover by (6):

$$m(\rho) \le |f(u_n)| = \max_t(|f(c_{1/n}(t) + h_{1/n}\gamma(t))|)$$
$$= I(c_{1/n} + h_{1/n}\gamma)$$
$$\le \inf I + \frac{1}{n} + Lh_0$$

and by (WPS) condition there exists u_0:

$$|f(u_0)| \ge m(\rho), \ \lambda(u_0) = 0 \ \text{i.e.} \ 0 \in \partial f(u_0). \ \square$$

Acknowledgement. This work is partially sponsored by the Bulgarian Ministry of Science and Education under Grant MM 68 / 91.

References

[1] Aubin J.P. L'analyze nonlineaire et ses motivations economique. Masson, Paris, 1984.

[2] Aubin J.P., Ekeland I. Applied Nonlinear Analysis. John Wiley & Sons, N.Y., 1984.

[3] Cakloviic L., Li S., Willem M. A note on Palais- Smale condition and coercivity. Diff. and Int. Eq. v.3, N 4, 799- 800 (1990).

[4] Chang K. Variational methods for non- differentiable functions and their applications . J. Math. Anal. Appl. 80, 102-129 (1981).

[5] Clarke F. Optimization and Nonsmooth Analysis. John Wiley & Sons, N.Y., 1983.

NONLINEAR NEUMANN PROBLEMS WITH CRITICAL EXPONENT IN SYMMETRICAL DOMAINS

ZHI-QIANG WANG

Department of Mathematics and Statistics,
Utah State University, Logan, UT 84322

Abstract. Consider the boundary value problem $(I)_\lambda$: $-\Delta u + \lambda u = |u|^{p-1}u$ in Ω, $\frac{\partial u}{\partial \nu} = 0$ on $\partial\Omega$, where Ω is a bounded smooth domain in $\mathbf{R}^N (N \geq 3)$ and $p = \frac{N+2}{N-2}$. We prove that, when Ω possesses suitable symmetries, for λ large $(I)_\lambda$ has many positive solutions having large energies and exhibiting multiple point concentrations.

INTRODUCTION

Following an extensive study of nonlinear Dirichlet problems in last decade, nonlinear Neumann problems have drawn attentions in last few years. The simplest model reads as follows

$$(I)_\lambda \begin{cases} -\Delta u + \lambda u = |u|^{p-1}u & \text{in } \Omega \\ \frac{\partial u}{\partial \nu} = 0 & \text{on } \partial\Omega \end{cases}$$

where $\Omega \subset R^N$ is a bounded domain with smooth boundary, $\lambda > 0$ is a constant, and ν is the unit outer normal to $\partial\Omega$. We consider critical exponent problems here, namely, $p = \frac{N+2}{N-2}$.

There have been several papers ([2], [3], [4], [5], [6], [15], [16], [17], [18], [20] and [23]) recently concerning the existence of positive solutions for problems like $(I)_\lambda$. Motivated by a result in [10] for Dirichlet problems and also results for the subcritical case of $(I)_\lambda$ of ours([21],[22]), in [23], by imposing some geometrical condition on $\partial\Omega$ we have established a multiplicity result for $(I)_\lambda$ which gives a lower bound on the number of nonconstant positive solutions in terms of the topology of the boundary $\partial\Omega$. More precisely, let $H(x)$ be the mean curvature with respect to the unit outward normal to $\partial\Omega$ at $x \in \partial\Omega$, and assume

$$\partial\Omega = \Gamma_1 \cup \Gamma_2 \cup \cdots \cup \Gamma_k$$

where each Γ_i is an $(N-1)$-manifold, disjoint with Γ_j, $j \neq i$. Define for $\delta > 0$,

$$\Gamma_i^\delta = \{x \in \Gamma_i | H(x) \geq \delta\}, \quad i = 1, 2, \ldots, k,$$

and

$$q_i = \sup_{\delta > 0} \text{cat}(\Gamma_i^\delta, \Gamma_i), \quad i = 1, 2, \ldots, k,$$

where $\text{cat}(\Gamma_i^\delta, \Gamma_i)$ denotes the Ljusternik-Schnirelman category of Γ_i^δ in Γ_i. We have the following results in [23]:

Theorem A([23]). For λ large, $(I)_\lambda$ possesses at least $q = \sum_{i=1}^k q_i$ distinct nonconstant positive solutions.

A special case is

Corollary([23]). Assume $H(x) > 0$, $\forall x \in \partial\Omega$. Then for λ large, $(I)_\lambda$ possesses at least $\text{cat}(\partial\Omega)$ nonconstant positive solutions.

Remark 1. The author has learned from G. Mancini that a similar result has been obtained recently by Adimurthi and Mancini([3]). In [6] some further results have been obtained relating the existence of positive solutions to the geometry of the boundary.

Remark 2. As was discussed in [17] and [18], problem $(I)_\lambda$ may be viewed as a prototype of pattern formation in biology and is related to the steady state problem for a chemotoctic aggregation model by Keller and Segel [13].

From the proof of Theorem A([23]), we can see that solutions given there are 'single-peaked' and of minimal energy type. Namely, let us define

(1) $$E_\lambda(u) = \int_\Omega (|\nabla u|^2 + \lambda u^2) dx, \quad u \in V_1(\Omega)$$

where

(2) $$V_1(\Omega) = \{u \in W^{1,2}(\Omega) | \int_\Omega |u|^{p+1} dx = 1\},$$

then all solutions obtained in Theorem A correspond to critical values of $E_\lambda(u)$ which are close to the absolute minimal value of E_λ, which we denote by m_λ and is defined by

$$m_\lambda = \min_{u \in V_1(\Omega)} E_\lambda(u).$$

And by the result in [6] all these solutions exhibit one point concentrations around somewhere on the boundary of Ω (See [6] for further details).

A problem which arises naturally along the lines of our study is to seek positive solutions of $(I)_\lambda$ having higher energy and multiple point concentrations. Considering that $(I)_\lambda$ is a model equation for pattern formation, we think it would be interesting and important to get the above question cleared up. In this note, we shall give some results in this direction for problem $(I)_\lambda$ in symmetrical domains.

A special example covered by our main result in this paper(Theorem B below) is when Ω is a solid torus in $\mathbf{R}^3$, obtained by rotating a disk in x, z plane centered at $(2, 0, 0)$ with radius 1 about the z-axis. The features here are : a) Ω is invariant under any rotation transformations about z-axis, especially, invariant under Z_k action given by

$$T(x, y, z) = (\cos\frac{2\pi}{k}x - \sin\frac{2\pi}{k}y, \sin\frac{2\pi}{k}x + \cos\frac{2\pi}{k}y, z),$$

or, when (x, y) is written as w in complex variable, by

(3) $$T(w, z) = (e^{\frac{2\pi}{k}} w, z);$$

and b) Ω does not contain any fix points of the above group action.

More generally, let k be a positive integer and let T be an isometric representation of the cyclic group Z_k of order k in $\mathbf{R}^N$. We make the following assumption: a).Ω is invariant under this action and b). $\overline{\Omega}$ does not contain any fix points of the action. Then, our main result in this note is the following

Theorem B. Assume that Ω has the properties mentioned as above. Then there exists $\lambda_* > 0$ s.t. $\forall \lambda \geq \lambda_*$, $(I)_\lambda$ possesses at least one positive solution $u_{\lambda,k}$ satisfying

$$(4) \qquad E_\lambda\left(\frac{u_{\lambda,k}}{\|u_{\lambda,k}\|_{L^{p+1}(\Omega)}}\right) = k^{\frac{2}{N}} 2^{-\frac{2}{N}} S + o(1) \quad \text{as } \lambda \to \infty,$$

where S is the best Sobolev constant.

As a corollary that covers the special example mentioned earlier, we have

Corollary. Let Ω be invariant under Z_k actions for any k and be disjoint with fix point sets of these actions. Then for any $k \geq 1$, there exists $\lambda_k > 0$ s.t. $\forall \lambda \geq \lambda_k$, $(I)_\lambda$ possesses at least one positive solution $u_{\lambda,k}$ satisfying

$$(5) \qquad E_\lambda\left(\frac{u_{\lambda,k}}{\|u_{\lambda,k}\|_{L^{p+1}(\Omega)}}\right) = k^{\frac{2}{N}} 2^{-\frac{2}{N}} S + o(1) \quad \text{as } \lambda \to \infty.$$

Remark 3. We shall see from the proof that $u_{\lambda,k}$ essentially has k-points concentration around the boundary $\partial\Omega$. By adopting the argument in [6] one should be able to prove that $u_{\lambda,k}$ has exactly k maxima points over $\overline{\Omega}$ which are all attained on $\partial\Omega$. Note that in [23] we have showed that

$$(6) \qquad m_\lambda \to 2^{-\frac{2}{N}} S, \quad \text{as } \lambda \to \infty.$$

Then from the above Corollary, for λ large enough $(I)_\lambda$ has many positive solutions staying at different energy levels.

THE PROOF OF THEOREM B

In order to establish the existence result we shall analyze the behavior of the minima of $E_\lambda(u)$ in subspaces $V_{1,k}(\Omega)$ defined by

$$(7) \qquad V_{1,k}(\Omega) = \{u \in V_1(\Omega) \mid u(Tx) = u(x) \ a.e. \text{in } \Omega\}$$

where T is the generator of the isometric representation of Z_k in $\mathbf{R}^N$ given above. Define

$$(8) \qquad m_{\lambda,k} = \min_{u \in V_{1,k}} E_\lambda(u).$$

To prove Theorem B, we need some asymptotic estimates which are extensions of some results in [23]. A key estimate will be

$$\lim_{\lambda \to \infty} m_{\lambda,k} = k^{\frac{2}{N}} 2^{-\frac{2}{N}} S.$$

Firstly, let S be the best Sobolev constant, i.e.

(9) $$S = \inf \left\{ \int_\Omega |\nabla u|^2 dx \mid u \in W_0^{1,2}(\Omega), \int_\Omega |u|^{p+1} dx = 1 \right\}$$

where $p = \frac{N+2}{N-2}$ as in $(I)_\lambda$ and will be fixed throughout this paper. For $\Omega = R^N$, S is achieved at a family of functions

$$u_{\epsilon,y}(x) = \frac{\alpha_{N,\epsilon}}{(\epsilon + |x-y|^2)^{\frac{N-2}{2}}} \quad \forall y \in R^N, \epsilon > 0$$

where $\alpha_{N,\epsilon}$ constant depending only on N and ϵ.

Proposition 1. Let $\lambda_n \to \infty$ and $u_n \in V_{1,k}(\Omega)$ satisfy

$$\overline{\lim_{n \to \infty}} E_{\lambda_n}(u_n) = A \leq k^{\frac{2}{N}} 2^{-\frac{2}{N}} S$$

then there exist a subsequence of u_n(still denoted by u_n) and $y_n \in \partial\Omega_n$, where $\Omega_n = \{x \in \mathbf{R}^N \mid \lambda_n^{-\frac{1}{2}} x \in \Omega\}$, for each $\epsilon > 0$ there exists $R > 0$ such that

$$\int_{B_R(y_n) \cap \Omega_n} |v_n|^{p+1} dx \geq \frac{1}{k} - \epsilon$$

where $v_n(x) = \lambda_n^{-\frac{N-2}{4}} u_n(\lambda_n^{-\frac{1}{2}} x)$.

This proposition can be proved using the same arguments as in [23] with the aid of concentration compactness lemma([14]).

Proposition 2. For $m_{\lambda,k}$ defined in (8),

(10) $$\lim_{\lambda \to \infty} m_{\lambda,k} = k^{\frac{2}{N}} 2^{-\frac{2}{N}} S.$$

Proof. By the result in [2] and [20] we know $m_{\lambda,1} < 2^{-\frac{2}{N}} S$, for any $\lambda > 0$. A lemma below actually will show that

$$m_{\lambda,k} < k^{\frac{2}{N}} 2^{-\frac{2}{N}} S.$$

To see that, let us define a family of functions $U_\epsilon(y) \in W^{1,2}(\mathbf{R}^N)$ by

$$U_\epsilon(y)(x) = \eta\left(\frac{|x-y|}{\rho}\right) \cdot \frac{1}{[\epsilon + (x-y)^2]^{\frac{N-2}{2}}} \quad \forall x \in \mathbf{R}^N$$

where $\eta(t) = 1$ for $t \leq 1$ and $\eta(t) = 0$ for $t \geq 2$. For $k \geq 1$ integer fixed, we choose $\rho > 0$ small enough and define a family of mappings $\varphi_\epsilon(\cdot) : \partial\Omega \to V_{1,k}(\Omega)$ by

$$\varphi_\epsilon(y)(x) = \frac{\sum_{i=1}^k U_\epsilon(T^i y)(x)}{\|\sum_{i=1}^k U_\epsilon(T^i y)\|_{L^{p+1}(\Omega)}}, \quad \forall y \in \partial\Omega, \quad x \in \Omega.$$

Lemma 1. For $\lambda > 0$ and $k \geq 1$ being fixed, if $H(y) > 0$, then

(11) $$E_\lambda(\varphi_\epsilon(y)) = k^{\frac{2}{N}} 2^{-\frac{2}{N}} S[1 - B_N H(y)\epsilon^{\frac{1}{2}} + o(\epsilon^{\frac{1}{2}})]$$

where B_N is a constant only depending on N, $H(y)$ is the mean curvature of $\partial\Omega$ at y.

The proof of this lemma follows from the calculations in [2] or [20].

Back to the proof of Proposition 2, if the conclusion is not true, there exist $\lambda_n \to \infty$, $u_n \in V_{1,k}(\Omega)$, and a constant $A < k^{\frac{2}{N}} 2^{-\frac{2}{N}} S$ such that

(12) $$\lim_{n\to\infty} \int_\Omega (|\nabla u_n|^2 + \lambda_n |u_n|^2) dx = A.$$

Rescaling u_n,

$$v_n(x) = \lambda_n^{-\frac{N-2}{4}} u_n(\lambda_n^{-\frac{1}{2}} x)$$

then $v_n \in V_{1,k}(\Omega_n)$, where $\Omega_n = \{x \in R^N \mid \lambda_n^{-\frac{1}{2}} x \in \Omega\}$. By (12)

$$\lim_{n\to\infty} \int_{\Omega_n} (|\nabla v_n|^2 + v_n^2) dx = A.$$

By embedding theorem, $W^{1,2}(\Omega_n) \hookrightarrow L^{p+1}(\Omega_n)$. Since Ω_n has uniform cone condition(see [1]), the embedding constant is independent of n and thus we may assume $A > 0$.

Now by Proposition 1, there exist $y_n \in \partial\Omega_n$ s. t. $\forall \epsilon > 0, \exists R > 0$

$$\int_{B_R(y_n) \cap \Omega_n} |v_n|^{p+1} dx \geq \frac{1}{k} - \epsilon.$$

For simplicity, assume $y_n = \lambda_n^{\frac{1}{2}} t_n$ where $t_n \in \partial\Omega$. We choose unitary matrix T_n such that $\tilde{\Omega}_n = T_n(\Omega_n - y_n)$ has y^N as inner normal direction of $\partial\tilde{\Omega}_n$ at θ. Then $T_n(\Omega - T^l y_n) = T_n T^l(\Omega_n - y_n)$ has the same property for $l = 1, 2, ..., k$, where T is the generator of the Z_k action. Now we introduce a diffeomorphism(See [18])which straightens a boundary portion near $\theta \in \partial(T_n(\Omega - t_n))$. Firstly, there exists a smooth function $\psi_n(x'), x' = (x^1, \cdots, x^{N-1})$, defined for $|x'| \leq a$ such that (i) $\psi(\theta) = \theta$ and $\nabla \psi_n(\theta) = \theta$; (ii) $\partial(T_n(\Omega - t_n)) \cap \mathcal{N} = \{(x', x^N) \mid x^N = \psi_n(x')\}$ and $T_n(\Omega - t_n) \cap \mathcal{N} = \{(x', x^N) \mid x^N > \psi_n(x')\}$, where $\mathcal{N}$ is a neighborhood of θ. Note that $\partial\Omega$ is compact, a can be chosen independent of n. Now for any $R' > 0$, $y \in \mathbf{R}^N$ with $|y| \leq R'$, define mappings $x = \Phi_n(y)$ with $\Phi_n(y) = ((\Phi_n)_1(y), \cdots, (\Phi_n)_N(y))$ by

(13) $\quad (\Phi_n)_j(y) = \begin{cases} y^j - y^N \frac{\partial \psi}{\partial x^j}(\lambda_n^{-\frac{1}{2}} y') & \text{for } j = 1, \ldots, N-1 \\ y^N + \lambda_n^{\frac{1}{2}} \psi(\lambda_n^{-\frac{1}{2}} y') & \text{for } j = N. \end{cases}$

which are well-defined for n large enough. And it is easy to verify that $D\Phi_n(y) \to Id_{N \times N}$ as $n \to \infty$ uniformly for $|y|$ bounded, where $Id_{N \times N}$ is $N \times N$ identity matrix. We write

(14) $\qquad\qquad D\Phi_n(y) = Id_{N \times N} + \mathcal{E}_n(y)$

with

(15) $\qquad\qquad \|\mathcal{E}_n(y)\| \to 0$ as $n \to \infty$, for $|y| \leq R'$.

Now let us back to the proof of Proposition 2, and recall that the compactness for v_n, i.e. $\forall \epsilon > 0, \exists R > 0, s.t.$

(16) $\qquad\qquad \int_{B_R(y_n) \cap \Omega_n} |v_n|^{p+1} dx \geq \frac{1}{k} - \epsilon,$

with $y_n \in \partial\Omega_n$ as above. For fixed $\epsilon > 0$, and $R > 0$, there is $R' > 0$, such that

(17) $$\Phi_n(B_{2R'}(\theta)) \supset B_{2R}(\theta) \cap (T_n(\Omega_n - y_n)).$$

Define for $l = 1, 2, .., k$,

(18) $$w_{n,l}(y) = \begin{cases} v_n(\Phi_n(y)) & \text{if } \Phi_n(y) \in B_{2R} \cap (T_n(\Omega_n - T^l y_n)) \\ v_n(\Phi_n(y', -y^N)) & \text{if } \Phi_n(y) \in B_{2R}(\theta) \backslash T_n(\Omega_n - T^l y_n) \end{cases}$$

then by setting

$$D_i = \Phi_n^{-1}(B_{iR}(\theta)) \subset B_{2R'}(\theta) \quad , \quad i = 1, 2,$$

one has

(19) $$w_{n,l}(y) = v_n(\Phi_n(y', |y^N|)) \qquad \text{for } y \in D_2.$$

Define

(20) $$\overline{w}_{n,l}(y) = \eta\left(\frac{|\Phi_n(y', |y^N|)|}{R}\right) w_{n,l}(y), \quad y \in R^N.$$

Then

$$\overline{w}_{n,l}(y) \in W^{1,2}(R^N),$$

and

$$\int_{R^N} |\overline{w}_{n,l}(x)|^{p+1} dx \geq \frac{2}{k} - \epsilon.$$

Finally, defining

$$\tilde{w}_n(x) = \sum_{l=1}^{k} \overline{w}_{n,l}(x)$$

then we have by (15)-(20)

(21) $$\int_{R^N} |\tilde{w}_n(y)|^{p+1} dy \geq \sum_{l=1}^{k} \int_{D_1} |w_{n,l}(y)|^{p+1} dy$$

$$\geq 2 \sum_{l=1}^{k} \int_{\{y \in D_1 \mid y^N > 0\}} |w_{n,l}(y)|^{p+1} dy$$

$$= 2 \sum_{l=1}^{k} \int_{B_R(T^l y_n) \cap \Omega_n} |v_n(x)|^{p+1} |D\Phi_n(y)| dx$$

$$\geq 2 - 2\epsilon \qquad \text{as} \quad n \to \infty.$$

By (13)-(15)

$$(22) \quad D(\Phi_n(y', |y^N|)) = \begin{cases} Id_{N\times N} + \mathcal{E}_n^+(y) & \text{for } y_N > 0 \\ Id_{(N-1)\times(N-1)} \dot{+}(-1) + \mathcal{E}_n^-(y) & \text{for } y_N < 0 \end{cases}$$

where like in (15), we also have

$$(23) \quad \|\mathcal{E}_n^\pm(y)\| \to 0 \qquad \text{as } n \to \infty \qquad \text{uniformly for } |y| \leq R'.$$

Now

$$(24) \int_{R_\pm^N} (|\nabla \tilde{w}_n|^2 + |\tilde{w}_n|^2)\,dy$$

$$\leq k \int_{R_\pm^N} \eta^2 \left(\frac{|\Phi_n(y)|}{R}\right)(|\nabla v_n(\Phi_n(y))|^2 + |v_n(\Phi_n(y))|^2)\,dy$$

$$+ \frac{4k}{R^2}\int_{R_\pm^N} |\nabla \eta|^2 |D\Phi_n(y)|^2 |v_n(\Phi_n(y))|^2\,dy+$$

$$k\int_{R_\pm^N} \eta^2 \|\mathcal{E}_n^\pm\|(|\nabla v_n(\Phi_n(y', |y^N|))|^2 + |v_n(\Phi_n(y', |y^N|))|^2)\,dy$$

$$\leq k\int_{R_\pm^N} \eta^2 \left(\frac{|\Phi_n(y)|}{R}\right)(|\nabla v_n(\Phi_n(y', |y^N|))|^2 + |v_n(\Phi_n(y', |y^N|))|^2)\,dy$$

$$+ \frac{C_1}{R^2} + C_2 \sup_{|y|\leq R'} \|\mathcal{E}_n^\pm\|$$

where C_1, C_2 constants independent of n. We may assume R large enough when we chose it such that

$$\frac{C_1}{R^2} < \frac{\epsilon}{2}.$$

Thus, by this and (22)-(24)

$$2\int_{\Omega_n} (|\nabla v_n|^2 + |v_n|^2)\,dx - \int_{R^N} (|\nabla \tilde{w}_n|^2 + |\tilde{w}_n|^2)\,dy$$

$$\geq 2k\int_{B_{2R(y_n)}\cap \Omega_n} (|\nabla v_n|^2 + |v_n|^2)\,dx - k\int_{D_2} (|\nabla \tilde{w}_n|^2 + |\tilde{w}_n|^2)\,dy$$

$$\geq 2k\int_{B_{2R}\cap\Omega_n} (|\nabla v_n|^2 + |v_n|^2)\,dx -$$

$$2k\int_{R_+^N \cap D_2} \eta^2 \left(\frac{|\Phi_n(y)|}{R}\right)(|\nabla v_n(\Phi_n(y))|^2 |D\Phi_n(y)|^2 + |v_n(\Phi_n(y))|^2)\,dy$$

$$-\frac{\epsilon}{2} - C_2 \sup_{|y| \leq R'} \|\mathcal{E}_n^{\pm}\|$$

$$\geq 2k \int_{B_{2R}(y_n) \cap \Omega_n} (|\nabla v_n|^2 + v_n^2) dx -$$

$$2k \int_{B_{2R}(0) \cap T_n(\Omega_n - y_n)} \eta^2 \left(\frac{|z|}{R}\right) (|\nabla v_n(T_n^{-1}(z+y_n))|^2 +$$

$$|v_n(T_n^{-1}(z+y_n))|^2) |D\Phi_n^{-1}| dz - \frac{\epsilon}{2} - C_2 \sup_{|y| \leq R'} \|\mathcal{E}_n^{\pm}\|$$

$$\geq 2k \int_{B_{2R}(y_n) \cap \Omega_n} \{|\nabla v_n(x))|^2 + |v_n(x))|^2\} \eta^2 \left(\frac{|x-y_n|}{R}\right) (1 - |D\Phi_n^{-1}|) dz$$

$$-\frac{\epsilon}{2} - C_2 \sup_{|y| \leq R'} \|\mathcal{E}_n^{\pm}\|$$

$$\geq -2\epsilon \qquad \text{as} \qquad n \to \infty$$

for by (22) we also have $D\Phi_n^{-1}(y) \to Id$, as $n \to \infty$, uniformly for $|y| \leq R'$.

Finally, by the last inequality together with (21) we can derive a contradiction as follows for $A < k^{\frac{2}{N}} 2^{-\frac{2}{N}} S$,

$$2A = \lim_{n \to \infty} 2 \int_{\Omega_n} (|\nabla v_n|^2 + v_n^2) dx$$

$$\geq \lim_{n \to \infty} \int_{R^N} (|\nabla \tilde{w}_n|^2 + |\tilde{w}_n|^2) dx - 2\epsilon$$

$$= \lim_{n \to \infty} \left(\sum_{l=1}^{k} \|\overline{w}_{n,l}\|_{L^{p+1}(R^N)}^{p+1} \right)^{\frac{2}{p+1}} \int_{R^N} \frac{|\nabla \tilde{w}_n|^2 + |\tilde{w}_n|^2}{\|\tilde{w}_n\|_{L^{p+1}(R^N)}^2} dx - 2\epsilon$$

$$\geq k(\frac{2}{k} - 2\epsilon)^{\frac{2}{p+1}} S - 2\epsilon$$

Letting $\epsilon \to 0$, one has

$$2A \geq k^{\frac{2}{N}} 2^{\frac{N-2}{N}} S$$

a contradiction.

The proof of Proposition 2 is complete. ∎

Next we turn to the compactness analysis for E_λ. Let c be a real number. Recall that E_λ satisfies $(P-S)_c$ condition in $V_{1,k}(\Omega)$ if whenever $\{u_n\} \in V_{1,k}(\Omega)$ satisfies $E_\lambda(u_n) \to c$, and $E'_\lambda(u_n)|_{V_{1,k}(\Omega)} \to 0$, then there exists a convergent subsequence of $\{u_n\}$.

Proposition 3. There exists a $\lambda_k > 0$, for any $\lambda \geq \lambda_k$, E_λ satisfies $(P-S)_c$ condition in $V_{1,k}(\Omega)$ for any $c < k^{\frac{2}{N}} 2^{-\frac{2}{N}} S$.

Lemma 2. Let $c > 0$ and $u \in V_{1,k}(\Omega)$ be a nontrivial solution of

(25)
$$\begin{cases} -\Delta u + \lambda u = c|u|^{p-1} u & \text{in } \Omega \\ \frac{\partial u}{\partial \nu} = 0 & \text{on } \partial\Omega \end{cases}$$

then

(26)
$$\|u\|_{L^{p+1}(\Omega)} \geq \left(\frac{m_{\lambda,k}}{c}\right)^{\frac{1}{p-1}}$$

Proof. Multiplying the equation (25) by u and integrate over Ω, one gets

$$\int_\Omega (|\nabla u|^2 + \lambda u^2) dx = c \int_\Omega |u|^{p+1} dx.$$

This implies

$$\|u\|_{L^{p+1}(\Omega)}^{p-1} = \frac{E_\lambda\left(\frac{u}{\|u\|_{p+1}}\right)}{c} \geq \frac{m_{\lambda,k}}{c}.$$

∎

The Proof of Proposition 3. Since $\|u_n\|_{W^{1,2}(\Omega)}$ is bounded, up to a subsequence we have $u_0 \in W^{1,2}(\Omega)$ s.t..

(27)
$$u_n \rightharpoonup u_0 \quad \text{in} \quad W^{1,2}(\Omega)$$

(28)
$$u_n \to u_0 \quad \text{in} \quad L^s(\Omega) \quad \forall\, s < \frac{2N}{N-2}.$$

We claim $u_0 \neq 0$. To see this, we need

Lemma 3 Let k be fixed. For any $\epsilon > 0$, there exists $C(\epsilon) > 0$ such that for all $u \in W^{1,2}(\Omega)$ with $u(Tx) = u(x)$ a.e. in Ω

$$\left(\int_\Omega |u|^{p+1} dx\right)^{\frac{2}{p+1}} \leq \left(k^{-\frac{2}{N}} 2^{\frac{2}{N}} S^{-1} + \epsilon\right) \int_\Omega |\nabla u|^2 dx + C(\epsilon) \int_\Omega |u|^2 dx.$$

This lemma is the symmetrical version of the following result in [7].

Lemma 4([7]). For any $\epsilon > 0$, there exists $C(\epsilon) > 0$ such that for all $u \in W^{1,2}(\Omega)$,

$$\left(\int_\Omega |u|^{p+1} dx\right)^{\frac{2}{p+1}} \leq \left(2^{\frac{2}{N}} S^{-1} + \epsilon\right) \int_\Omega |\nabla u|^2 dx + C(\epsilon) \int_\Omega |u|^2 dx.$$

The Proof of Lemma 3. Using Z_k symmetry, we slice Ω into k equal subdomains, say, O_i, $i = 1, 2, ...k$. and apply lemma 4 to each O_i

$$\left(\int_{O_i} |u|^{p+1} dx\right)^{\frac{2}{p+1}} \leq \left(2^{\frac{2}{N}} S^{-1} + \epsilon\right) \int_{O_i} |\nabla u|^2 dx + C(\epsilon) \int_{O_i} |u|^2 dx.$$

And we add above k formulas together and reset the constants there. Then Lemma 3 follows from

$$\left(\int_\Omega |u|^{p+1} dx\right)^{\frac{2}{p+1}} = \left(\sum_{i=1}^k \int_{O_i} |u|^{p+1} dx\right)^{\frac{2}{p+1}} \leq \sum_{i=1}^k \left(\int_{O_i} |u|^{p+1} dx\right)^{\frac{2}{p+1}}.$$

∎

Applying Lemma 3 to u_n, by (28) one has for any $\epsilon > 0$

$$1 \leq (k^{-\frac{2}{N}} 2^{\frac{2}{N}} S + \epsilon) \lim_{n \to \infty} \int_\Omega |\nabla u_n|^2 dx.$$

Now, if $u_0 \equiv 0$, $\lim_{n \to \infty} \int_\Omega |\nabla u_n|^2 dx \neq 0$.

By Lemma 3, and (28), for $\epsilon > 0$ fixed,

$$\lim_{n \to \infty} \int_\Omega |\nabla u_n|^2 dx = \lim_{n \to \infty} E_\lambda(u_n)$$

$$\leq c \left(k^{-\frac{2}{N}} 2^{\frac{2}{N}} S^{-1} + \epsilon\right) \lim_{n \to \infty} \int_\Omega |\nabla u_n|^2 dx$$

Therefore,

$$1 \leq c \left(k^{-\frac{2}{N}} 2^{\frac{2}{N}} S^{-1} + \epsilon\right), \forall \epsilon > 0.$$

Letting $\epsilon \to 0$,

$$c \geq k^{\frac{2}{N}} 2^{-\frac{2}{N}} S,$$

a contradiction. So $u_0 \neq 0$.

By $E'_\lambda(u_n)|_{V_{1,k}(\Omega)} \to 0$ and symmetric criticality principle([19]), for any $\varphi \in C^\infty(\Omega)$,

(29) $$\int_\Omega (\nabla u_n \nabla \varphi + \lambda u_n \varphi) dx - c \int_\Omega |u_n|^{p-1} u_n \varphi dx \to 0 \quad n \to \infty.$$

By (27)-(29), we get
$$\int_\Omega (\nabla u_0 \nabla \varphi + \lambda u_0 \varphi) dx - c \int_\Omega |u_0|^{p-1} v_0 \varphi dx = 0,$$

i.e. u_0 is a solution of (25). By Lemma 2,
$$\|u_0\|_{L^{p+1}(\Omega)}^{p-1} \geq \frac{m_{\lambda,k}}{c}.$$

Define $v_n(x) = u_n(x) - u_0(x)$, then

(30)
$$\int_\Omega (|\nabla v_n|^2 + \lambda |v_n|^2) dx$$
$$= \int_\Omega (|\nabla u_n|^2 + \lambda |u_n|^2) dx + \int_\Omega (|\nabla u_0|^2 + \lambda |u_0|^2) dx$$
$$- 2 \int_\Omega (\nabla u_n \nabla u_0 + \lambda u_n u_0) dx$$
$$= c - \int_\Omega (|\nabla u_0|^2 + \lambda |u_0|^2) dx + o(1)$$
$$= c - c\|u_0\|_{L^{p+1}(\Omega)}^{p+1} + o(1)$$
$$= (1 - \|u_0\|_{L^{p+1}(\Omega)}^{p+1}) c + o(1).$$

By Proposition 2, $m_{\lambda,k} \to k^{\frac{2}{N}} 2^{-\frac{2}{N}} S$, as $n \to \infty$. Then there exists $\lambda_k > 0$ such that for $\lambda \geq \lambda_k$,
$$\left[1 - \left(\frac{m_{\lambda,k}}{k^{\frac{2}{k}} 2^{-\frac{2}{N}} S}\right)^{\frac{N}{2}}\right]^{\frac{2}{N}} k^{\frac{2}{N}} 2^{-\frac{2}{N}} S < m_{\lambda,k}.$$

Now we claim $\|u_0\|_{L^{p+1}(\Omega)} = 1$. Then (30) shows
$$\int_\Omega |\nabla(u_n - u_0)|^2 dx \to 0,$$

i.e. $u_n \to u_0$ in $W^{1,2}(\Omega)$.

To prove the claim, firstly, by Brezis-Lieb Lemma([11]),
$$1 = \|u_n\|_{L^{p+1}(\Omega)}^{p+1} = \|u_0\|_{L^{p+1}(\Omega)}^{p+1} + \|v_n\|_{L^{p+1}(\Omega)}^{p+1} + o(1).$$

Then, if $\|u_0\|_{L^{p+1}(\Omega)} \neq 1$,

$$E_\lambda \left(\frac{v_n}{\|v_n\|_{L^{p+1}(\Omega)}} \right) = \frac{(1 - \|u_0\|_{L^{p+1}(\Omega)}^{p+1})c + o(1)}{(1 - \|u_0\|_{L^{p+1}(\Omega)}^{p+1})^{\frac{2}{p+1}} + o(1)}$$

$$= \left(1 - \|u_0\|_{L^{p+1}(\Omega)}^{p+1} \right)^{\frac{2}{N}} c + o(1)$$

$$\leq \left[1 - \left(\frac{m_{\lambda,k}}{c} \right)^{\frac{N}{2}} \right]^{\frac{2}{N}} c + o(1)$$

$$\leq \left[1 - \left(\frac{m_{\lambda,k}}{k^{\frac{2}{N}} 2^{-\frac{2}{N}} S} \right)^{\frac{N}{2}} \right]^{\frac{2}{N}} c + o(1)$$

$$< m_{\lambda,k} + o(1) \qquad \text{for} \qquad \lambda \geq \lambda_k.$$

A contradiction with the definition of $m_{\lambda,k}$. The proof of Proposition 3 is complete. ∎

Finally, we need

Lemma 5. Let u be a critical point of $E_{\lambda,k}$ in $V_{1,k}(\Omega)$ with $E_{\lambda,k}(u) < 2^{\frac{2}{N}} m_{\lambda,k}$, then u does not change sign in Ω.

Proof. Note that when u satisfies $u(Tx) = u(x)$, $u_+(x) = \max\{u(x), 0\}$ and $u_-(x) = \min\{u(x), 0\}$ also satisfy the property. Then the proof follows step by step the proof of Lemma 3.6 in [23]. ∎

Now, **the proof of Theorem B.** Firstly, by the symmetric criticality principle[11], any critical point of E_λ in $V_{1,k}(\Omega)$ is a critical point of E_λ in $V_1(\Omega)$ and therefore is a solution of $(I)_\lambda$. Then the proof follows from Proposition 1, 2, Lemma 1 and 5. ∎

Remark 4. We may consider more general boundary conditions. Let Γ' and Γ be disjoint submanifolds of $\partial\Omega$ such that $\partial\Omega = \Gamma' \cup \Gamma$, and consider

$$(I)'_\lambda \begin{cases} -\Delta u + \lambda u = |u|^{p-1}u & \text{in } \Omega \\ \frac{\partial u}{\partial \nu} = 0 & \text{on } \Gamma \\ u = 0 & \text{on } \Gamma' \end{cases}$$

Theorem C. Assume that there exists $y \in \Gamma$ s.t. $H(y) > 0$. If Ω is invariant under Z_k symmetry and every Z_k orbit in $\overline{\Omega}$ contains exactly k points, then for λ large enough, $(I)'_\lambda$ possesses a positive solution $u_{\lambda,k}$ with the property given in (4).

Acknowledgment. This research was partially supported by an NSF grant (DMS-9201283) and a faculty research grant at Utah State University (SM-11122).

REFERENCES

[1] R. Adams, "Sobolev Spaces," Academic Press, New York, 1975.

[2] Adimurthi & G. Mancini, The Neumann problem for elliptic equations with critical non-linearity, Estratto da Nonlinear Analysis, Scuola Normale Superiore, Pisa (1991), p9-25.

[3] Adimurthi & G. Mancini, Effect of geometry and topology of the boundary in the critical Neumann problem, preprint.

[4] Adimurthi & S.L. Yadava, Critical Sobolev exponent problem in $\mathbf{R}^N (n \geq 4)$ with Neumann boundary condition, Proc. Indian Acad. Sci. (Math. Sci.) **100**, No. 3, Dec. 1990, p. 275-284.

[5] Adimurthi & S.L. Yadava, Existence and nonexistence of positive radial solutions of Neumann problems with Critical Sobolev exponents, Arch. Rational Mech. Anal. **115** (1991), p 275-296.

[6] Adimurthi, F. Pacella & S.L. Yadava, Interaction between the geometry of the boundary and positive solutions of a semilinear Neumann problem with critical nonlinearity, preprint.

[7] T. Aubin, Nonlinear analysis on manifold, Monge-Ampere equations, Springer-Verlag, (1982).

[8] H. Brézis, Nonlinear elliptic equations involving the critical Sobolev exponent-Survey and Perspectives in Directions in P.D.E. Eds Grandall, Rabinowitz and Turner), (1987) p.17-36.

[9] A. Bahri & J.M. Coron, On a nonlinear elliptic equation involving the Sobolev exponent: the effect of the topology of the domain, Comm. Pure Appl. Math. **41** (1988), 253–294.

[10] V. Benci & G. Cerami, The effect of the domain topology on the number of positive solutions of nonlinear elliptic problems, Arch. Rational Mech. Anal. **114** No 1, 1991, p 79-93.

[11] H. Brézis & E. Lib, A relation between pointwise convergence of functions and convergence of functionals, Proc. Am. Math. Soc. **88** (1983) p.486-490.

[12] H. Brézis & L. Nirenberg, Positive solutions of nonlinear elliptic equations involving critical exponents, Comm. Pure Appl. Math. **36** (1983) p.437-477.

[13] E.F. Keller & L.A. Segel, Initiation of slime model aggregation viewed as an instability, J. Theor. Biol. **26** (1970), 399–415.

[14] P.L. Lions, The concentration-compactness principle in the calculus of variations, The locally compact case, Part 1 and Part 2, Ann. Inst. H. Poincaré Anal. Nonlinéaire **1** (1984), 109–145, 223–283.

[15] P.L. Lions, F. Pacella & M. Tricarico, Best constants in Sobolev inequalities for functions vanishing on some parts of the boundary and related questions, Indiana Univ. Math. J. **37** (1988). p.301-324.

[16] C.-S. Lin & W.-M. Ni, On the diffusion coefficient of a semilinear Neumann problem, in "Calculus of Variations and Partial Differential Equations,"(S. Hildebrandt, D. Kinderlehrer and M. Miranda, Ed.) 160-174, Lecture Notes in Math. 1340, Springer-Verlag 1988.

[17] C.-S. Lin, W.-M. Ni & I. Takagi, Large amplitude stationary solutions to a chemotaxis system, J. Diff. Euqa. **72** (1988), 1-27.

[18] W.-M. Ni & I. Takagi, On the shape of least-energy solutions to a semilinear Neumann problem, Comm. Pure Appl. Math., **45**, 1991, No 7, 819-851.

[19] R. Palais, The principle of symmetric criticality. Comm. Math. Phys. **69**, 1979, 19-30.

[20] X.-J. Wang, Neumann problems of semilinear elliptic equations involving critical Sobolev exponents, J. Diff. Equ. **93** (1991) p.283-310.

[21] Z.-Q. Wang, On the existence of multiple, single-peaked solutions of a semilinear Neumann problem, To appear in Arch. Rat. Mech. Anal..

[22] Z.-Q. Wang, On the existence of positive solutions for semilinear Neumann problems in exterior domains, To appear in Comm. on P.D.E..

[23] Z.-Q. Wang, The effect of the domain geometry on the number of positive solutions of Neumann problems with critical exponents, preprint.

SOME RESULTS ON THE GEODESIC CONNECTEDNESS OF LORENTZIAN MANIFOLDS

ANTONIO MASIELLO
Dipartimento di Matematica-Politecnico di Bari-Via Orabona 4
70125 BARI (ITALY)

1. INTRODUCTION

A Lorentzian manifold is a couple $(M, \langle \cdot, \cdot \rangle_L)$, where M is a smooth connected finite dimensional manifold, and $\langle \cdot, \cdot \rangle_L$ is a Lorentzian metric on M, i.e. a smooth (0,2) symmetric tensor field, such that for every $z \in M$, $\langle \cdot, \cdot \rangle_L$ defines a nondegenerate bilinear form on the tangent space T_zM, having index 1 (see also [ON]).

The study of the geometry of a Lorentzian manifold has been a great interest in this century, because the models of space-time in General Relativity are 4-dimensional Lorentzian manifolds.

In order to study the global geometry of a Lorentzian manifold, geodesic curves play an important role. We recall that a smooth curve $\gamma:]a,b[\longrightarrow M$ is said geodesic if

$$\nabla_s \dot\gamma = 0,$$

where $\dot\gamma$ is the tangent vector field along γ, and $\nabla_s \dot\gamma$ is the covariant derivative of $\dot\gamma$ along γ induced by the metric $\langle \cdot, \cdot \rangle_L$.

It is well known that if γ is a geodesic, there exists a constant $E(\gamma)$ such that

$$E(\gamma) = \langle \dot\gamma(s), \dot\gamma(s) \rangle_L \quad \forall s \in]a,b[.$$

Then γ is said

timelike   if $E(\gamma) < 0$;

lightlike   if $E(\gamma) = 0$;

spacelike,   if $E(\gamma) > 0$.

This classification is called the *causal character of geodesics.*

The terminology of the causal character comes from General Relativity. In particular a timelike geodesic represents the trajectory of a free falling particle in a space-time. Lightlike geodesics represent the light rays. The spacelike geodesics have no physical meaning, because travelling on it, a particle should be faster than light. However, they are interesting in the study of the global geometry of a Lorentzian manifold.

A global problem which naturally arises is the geodesic connectedness. A Lorentzian geometry (and in general a Semiriemannian manifold) is said to be geodesically connected if every couple of points may be joined by a geodesic.

The study of the geodesic connectedness has a satisfactory answer in the Riemannian case. Indeed, by the Hopf-Rinow theorem

Every complete Riemannian manifold is geodesically connected.

In the Lorentzian case this result has not a reasonable extension, since a distance is not naturally defined (as in the Riemannian case), because of the indefinitess of the metric.

It is well known that by Hopf-Rinow Theorem, a Riemannian manifold is complete iff it is geodesically complete, i.e. every geodesic can be extended to a geodesic defined in $\mathbb{R}$. Hence, every geodesically complete Riemannian manifold is geodesically connected.

This fact is not true in the Lorentzian case. For instance, there exist compact Lorentzian manifolds which are not geodesically connected (see for instance [BP]). Another classical example of a geodesically complete but not geodesically connected Lorentzian manifold is the Anti de Sitter space-time (see [Pe]) *M*

$=]-\frac{\pi}{2},\frac{\pi}{2}[\times \mathbb{R}$, equipped with the Lorentzian metric

$$ds^2 = \frac{(dx)^2-(dt)^2}{\cos^2 x}.$$

Some results on the geodesic connectedeness have been obtained, using variational methods, in [BF1, BF2, BFG1] for stationary Lorentzian manifolds, and [BFG2,BFG3,GM] for stationary Lorentzian manifolds with boundary (see [LL] for the physical meaning of a stationary space-time and [BFM1] for a survey of these results).

A result on the geodesic connectedeness of non necessarily stationary Lorentzian manifolds was obtained in [BFM2] for product manifolds $M = M_0 \times \mathbb{R}$.

Other results has been obtained in [Gi], where it is studied the geodesic connectedeness of a strip $M = M_0 \times]a,b[\subseteq M_0 \times \mathbb{R}$, and the metric is singular on the boundary $\partial M = M_0 \times \{a\} \cup M_0 \times \{b\}$.

In this paper we consider a strip $M = M_0 \times]a,b[$, and make different assumptions from [Gi], in order to obtain the geodesic connectedness of the strip. In particular, we consider a product manifold

$$\tilde{M} = M_0 \times \mathbb{R}, \qquad (1.1)$$

equipped with a Lorentzian metric of the form

$$\langle \zeta,\zeta \rangle_L = \langle (\xi,\tau),(\xi,\tau) \rangle = \langle \alpha(z)\xi,\xi \rangle - \beta(z)\tau^2, \qquad (1.2)$$

for all $z = (x,t) \in \tilde{M}$, $\zeta = (\xi,\tau) \in T_z \tilde{M} \equiv T_x M_0 \times \mathbb{R}$, where $\langle \cdot,\cdot \rangle$ is a Riemannian metric on M_0, $\alpha(z) = \alpha(x,t)$ is a positive linear operator on $T_x M_0$ which depends smoothly on z, and β is a smooth positive function on M.

These assumptions are not too restrictive. Indeed Geroch has proved that every globally hyperbolic Lorentzian manifold satisfies (1.1) and (1.2) (see [Ge,U] also for the definition of globally hyperbolic Lorentzian metric).

We have the following theorem about the geodesic connectedness of a strip $M = M_0 \times]a,b[$.

THEOREM 1.1 - *Let $(\tilde{M}, \langle \cdot, \cdot \rangle_L)$ be a Lorentzian manifold which satisfies (1.1) and (1.2), and $M = M_0 \times]a,b[$. Assume that:*

A_1) $(M_0, \langle \cdot, \cdot \rangle)$ *is a complete Riemannian manifold;*

A_2) *there exists a constant $\lambda > 0$, such that for every $z = (x,t) \in M$ and for every $\xi \in T_x M_0$:*

$$\lambda \langle \xi, \xi \rangle \leq \langle \alpha(x,t)\xi, \xi \rangle;$$

A_3) *there exist two constants $0 < \nu \leq C$, such that for every $z \in M$:*

$$\nu \leq \beta(z) \leq C;$$

we set for simplicity $C = 1$.

A_4)
$$\sup \left\{ |\beta_t(z)|, \|\alpha_t(z)\|, z \in \tilde{M} \right\} = L < +\infty,$$

where $\beta_t(z)$ and $\alpha_t(z)$ denote the partial derivatives with respect to t of β and α;

A_5) *there exists $\eta > 0$, such that for every $x \in M_0$, $\xi \in T_x M_0$:*

i) $\langle \alpha_t(x,t)\xi, \xi \rangle \leq 0,$ $\forall t \in]b - \eta, b[$,

ii) $\langle \alpha_t(x,t)\xi, \xi \rangle \geq 0.$ $\forall t \in]a, a + \eta[$

Then M, equipped with (1.2), is geodesically connected.

REMARK 1.2 - Assumptions A_2) and A_3) require some properties of nondegeneration of α and β. Assumption A_4) requires a bound on the growth of α and β.

REMARK 1.3 - Assumption A_5), which may appear only technical, play a fundamental role to prove Theorem 4.1, and has a geometrical meaning because it implies that the strip M has convex boundary $\partial M = M_0 \times \{a\} \cup M_0 \times \{b\}$, according to the definition given in [BFG2] (see [M] for a detailed discussion of the convexity of the boundary of a strip). In [BFG2,GM1] it has been proved the geodesic connectedness of stationary Lorentzian manifolds with convex boundary. Theorem 1.1 may be considered an extension of the results of those papers to nonstationary Lorentzian metrics.

REMARK 1.4 - Assumption A_5) can be weakened, using some estimates more delicate than those of section 4 (see [BFM2] and [Ma]).

From Theorem 1.1 we deduce the following Theorem on the geodesic connectedness of the whole manifold $\tilde{M} = M_0 \times \mathbb{R}$.

THEOREM 1.5 - *Assume that* A_1) *holds, and there exist functions* $\lambda(t)$, $\nu(t)$, $C(t)$, $M(t)$ $:\mathbb{R} \longrightarrow \mathbb{R}^+$, *such that for every* $x \in M_0$, $\xi \in T_x M_0$, $t \in \mathbb{R}$:

B_2) $\qquad\qquad\qquad \lambda(t)\langle\xi,\xi\rangle \leq \langle\alpha(x,t)\xi,\xi\rangle;$

B_3) $\qquad\qquad\qquad \nu(t) \leq \beta(x,t) \leq C(t);$

B_4) $\qquad\qquad\qquad \left\{|\beta_t(x,t)|, \|\alpha_t(x,t)\|\right\} < M(t);$

B_5) *there exists two real sequences* $(a_h)_{h\in\mathbb{N}}$, $(b_k)_{k\in\mathbb{N}}$, *such that* $(a_h)_{h\in\mathbb{N}}$ *converges to* $-\infty$, $(b_k)_{k\in\mathbb{N}}$ *converges to* $+\infty$, *and for every* $k,h \in \mathbb{N}$, *the submanifold* $M_{h,k} = M_0 \times]a_h,b_k[$ *satisfies assumption* A_5) *of Theorem 1.1.*

Then $\tilde{M}$ is geodesically connected.

REMARK 1.6 - Notice that if M_0 is a compact manifold, assumptions A_1) and B_2)-B_4) are always satisfied.

REMARK 1.7 - The statement of Remark 1.4 holds also for Theorem 1.5.

REMARK 1.8 - Assumption B_5) cannot be removed. Indeed consider on the manifold $\mathbb{R} \times \mathbb{R}$, the Lorentzian metric

$$ds^2 = \cosh^2 t\, dx^2 - dt^2. \tag{1.3}$$

(1.3) satifies assumptions A_1) and B_2)-B_4), but not B_5). Moreover, $\mathbb{R} \times \mathbb{R}$, equipped with (1.3), is not geodesically connected. Indeed it is isometric to the Lorentzian manifold $\mathbb{R} \times]-\frac{\pi}{2},\frac{\pi}{2}[$, equipped with the metric

$$ds^2 = \frac{dx^2 - d\tau^2}{\cos^2 \tau}, \tag{1.4}$$

$x \in \mathbb{R}$ and $\tau \in]-\frac{\pi}{2},\frac{\pi}{2}[$, which is the Anti de Sitter space-time with space and time changed. This manifold is not geodesically connected as the Anti de Sitter space-time.

REMARK 1.9 - Suppose that assumptions A_1)-A_5) hold. Moreover, suppose that M_0 is not contractible in itself. Then, for every couple of points of M, there exists a sequence $(\gamma_m)_{m \in \mathbb{N}}$ of geodesics joining them, such that

$$\lim_{m \to \infty} E(\gamma_m) = +\infty.$$

This fact can be proved using the estimates of Section 4 and the techniques developped in [BF2] in order to get infinitely many geodesics joining two points of a stationary Lorentzian manifold.

2. ANALYTICAL PRELIMINAIRES

In this section we introduce the functional framework in order to prove Theorem 1.1.

Let $(\tilde{M},<\cdot,\cdot>_L)$ be a Lorentzian manifold which satisfies (1.1) and (1.2). By the Nash imbedding theorem (see [Na]), we can assume that M_0 is a submanifold of the euclidean space $\mathbb{R}^N$, with $N = \frac{n}{2}(n+1)(3n+11)$, $n = \dim M_0$, and the Riemannian metric on M_0 $<\cdot,\cdot>$ is the Euclidean one.

Now we introduce the functional manifolds in which we shall work. We set $I = [0,1]$.

Let $z_0 = (x_0,t_0)$ and $z_1 = (x_1,t_1)$ be two points of the strip

$M = M_0 \times \,]a,b[\, \subseteq \tilde M$. It is well known that the geodesics joining z_0 and z_1 in $\tilde M$ are the critical points of the action integral

$$f(z) = f(x,t) = \frac{1}{2}\int_0^1 <\dot z(s),\dot z(s)>_L ds =$$
$$= \frac{1}{2}\int_0^1 [<\alpha(z)\dot x,\dot x> - \beta(z)\dot t^2]ds, \qquad (2.1)$$

defined on the Hilbert manifold

$$Z = \Omega^1 \times H^{1,2}(t_0,t_1;]a,b[), \qquad (2.2)$$

where

$$\Omega^1 = \Omega^1(M_0,x_0,x_1) =$$
$$\{x:I \to M_0 | x \in H^{1,2}(I,\mathbb{R}^N), x(0) = x_0, x(1) = x_1\}, \qquad (2.3)$$

and

$$H^{1,2}(t_0,t_1;]a,b[) =$$
$$\{t \in H^{1,2}(I,]a,b[) | t(0) = t_0, t(1) = t_1\}. \qquad (2.4)$$

We recall that for every $k \in \mathbb{N}$, $H^{1,2}(I,\mathbb{R}^k)$ is the Sobolev space of absolutely continous curves in $\mathbb{R}^k$, whose derivative is square summable. It is a Hilbert space with norm

$$\|x\|_1^2 = \|x\|^2 + \|\dot x\|^2,$$

where $\dot x$ is the derivative of x, and $\|\cdot\|$ is the usual norm of $L^2(I,\mathbb{R}^k)$ (in this paper we always shall denote by $\|\cdot\|$ the L^2-norm).

By virtue of $A_1)$, see for instance [Pa], Ω^1 is a complete Riemannian submanifold of $H^{1,2}(I,\mathbb{R}^N)$. For every $x \in \Omega^1$, the tangent space is

$$T_x\Omega^1 = \{\xi:I \to \mathbb{R}^N | \xi \in H_0^{1,2}(I,\mathbb{R}^N), \xi(s) \in T_{x(s)}M_0 \; \forall s \in I\},$$

where, for every $k \in \mathbb{N}$,

$$H_0^{1,2}(I,\mathbb{R}^k) = \{\xi \in H^{1,2}(I,\mathbb{R}^k) | \xi(0) = \xi(1) = 0\}.$$

Moreover $H^{1,2}(t_0,t_1;]a,b[)$ is an open subset of

$$H^{1,2}(t_0,t_1) = \{t \in H^{1,2}(I,\mathbb{R}) | t(0) = t_0, \ t(1) = t_1\}.$$

$H^{1,2}(t_0,t_1)$ is a closed linear submanifold of $H^{1,2}(I,\mathbb{R})$. Indeed, let

$$\bar{t}(s) = t_0 + s(t_1-t_0)$$

be the segment joining t_0 and t_1, then

$$H^{1,2}(t_0,t_1) = \bar{t} + H_0^{1,2}(I,\mathbb{R}).$$

Hence Z is the manifold of the curves in $H^{1,2}(I,\mathbb{R}^{N+1})$ joining z_0 and z_1 in $\tilde{M}$. The tangent space to Z of a curve $z = (x,t)$ is given by

$$T_z Z = T_x \Omega^1 \times H_0^{1,2}(I,\mathbb{R}).$$

The action integral (2.1) is strongly unbounded both from below and from above, and its critical points have Morse index $+\infty$. Moreover it does not satisfy the compactness condition of Palais-Smale (P.S.), beacuse of the lack of compactness of the manifold Z (indeed Z is not a complete manifold), and the dependence on the "time variable" of the coefficients α and β of the metric (see [BF2] for the (P.S.) for stationary metrics).

In order to overcome these problems, we use a Galerkin approximation argument for the indefinitess of the action integral, and a penalization argument for the lack of (P.S.).

Now we introduce the penalization argument. Let $\varphi:\mathbb{R} \longrightarrow \mathbb{R}$ be the function defined by $\varphi(t) = (t-a)(b-t)$, and let for every $z = (x,t) \in M$, $\phi(z) = \varphi(t)$. Then we have:

$$M = \{z \in \tilde{M} | \phi(z) > 0\},$$

$$\partial M = M_0 \times \{a\} \cup M_0 \times \{b\} = \{z \in \tilde{M} \mid \phi(z) = 0\},$$

$$\nabla \phi(z) \neq 0, \text{ for every } z \in \partial \tilde{M},$$

where $\nabla\phi(z)$ is the gradient of ϕ in z, with respect to the metric $<\cdot,\cdot>_L$ (notice that since ϕ depends only on the variable t, $\nabla\phi(z)$ dependes essentially on $\varphi'(t)$).

For every $\delta > 0$, we consider the penalized functional $f_\delta : Z \longrightarrow \mathbb{R}$, such that for every $z = (x,t) \in Z$

$$f_\delta(x,t) = f(z) + \frac{\delta}{4}\left[\|\dot{x}\|^4 - \|\dot{t}\|^4\right] - \delta \int_0^1 \frac{1}{\varphi^2(t)} ds, \qquad (2.5)$$

For every $\delta > 0$, the penalized functional f_δ satisfies (P.S.) (see section 3).

Now we introduce the Galerkin approximation argument. For every $k \in \mathbb{N}$, let

$$H^{1,2}_{0,k} = \text{span}\left\{ \sin(\pi l s),\ 1 \leq l \leq k \right\}.$$

$H^{1,2}_{0,k}$ is a finite dimensional subspace of $H^{1,2}_0(I,\mathbb{R})$.

Moreover, we set

$$H^{1,2}_k(t_0,t_1) = \bar{t} + H^{1,2}_{0,k},$$

which is a finite dimensional submanifold of $H^{1,2}(t_0,t_1)$.

Finally we set

$$H^{1,2}_k(t_0,t_1;]a,b[) = H^{1,2}_k(t_0,t_1) \cap H^{1,2}(t_0,t_1;]a,b[),$$

and

$$Z_k = \Omega^1 \times H^{1,2}_k(t_0,t_1;]a,b[). \qquad (2.6)$$

Notice that Z_k is a submanifold of Z, whose "time" component has finite dimension.

We state now a critical point point Theorem, which will be applied to find a critical point of the penalized functional f_δ.

It is a generalization of the well known Saddle Point Theorem of Rabinowitz (see [Ra]). First we recall the Palais-Smale condition (P.S.) for smooth functionals.

DEFINITION 2.1 - Let $f: X \longrightarrow \mathbb{R}$ be a smooth functional defined on the Riemannian manifold X. F satisfies (P.S.) if every sequence $(x_k)_{k \in \mathbb{N}}$ such that

$$\{f(x_k)\}_{k \in \mathbb{N}} \text{ is bounded,}$$

$$\|f'(x_k)\|_{x_k} \xrightarrow[k \to \infty]{} 0,$$

contains a converging subsequence in X. Here, for every $x \in X$, $\|\cdot\|_x$ denotes the norm induced on $T_x X$ by the Riemannian metric.

Let $\bar{x} \in \Omega^1$ fixed and $\bar{t}$ the segment joining t_0 and t_1. We set

$$S = \{(x, \bar{t}), x \in \Omega^1\} \subseteq Z. \tag{2.7}$$

Moreover, for every $k \in \mathbb{N}$, $R > 0$, we set

$$Q(R) = \{(\bar{x}, t) \in Z \mid \|t - \bar{t}\|_1 < R\}, \tag{2.8}$$

$$Q_k(R) = \{(\bar{x}, t) \in Z_k \mid \|t - \bar{t}\|_1 < R\}. \tag{2.9}$$

Notice that Q and Q_k are Hilbertian submanifolds of Z and Z_k respectively, whose boundaries are

$$\partial Q(R) = \{(\bar{x}, t) \in Z \mid \|t - \bar{t}\|_1 = R\}, \tag{2.10}$$

$$\partial Q_k(R) = \{(\bar{x}, t) \in Z_k \mid \|t - \bar{t}\|_1 = R\} \tag{2.11}$$

We have the following slight variant of the saddle point Theorem (see [Ra, BF1]).

THEOREM 2.2 - Let $I: Z \longrightarrow \mathbb{R}$ be a C^1 functional and I_k the restriction of I to Z_k, where Z and Z_k are defined in (2.2) and (2.6) respectively. Assume that:

(i) *for every* $a < b \in \mathbb{R}$, *the subset*

$$f_a^b = \{z \in Z | a \le f(z) \le b\}$$

is a complete metric subspace of Z;

(ii) I_k *satisfies (P.S.) condition for every* $k \in \mathbb{N}$;

(iii) $\exists R > 0$, *such that:*

a) $\inf I(S) > -\infty$;

b) $\sup I(Q(R)) < +\infty$;

c) $\sup I(\partial Q(R)) < \inf I(S)$.

Define, for every $k \in \mathbb{N}$:

$$c_k = \inf_{h \in \Gamma_k} \sup I(h(Q_k(R))),$$

where

$$\Gamma_k = \left\{ h \in C(Z_k, Z_k) | h(z) = z \text{ for every } z \in \partial Q_k(R) \right\}.$$

Then every c_k is well defined, $c_k \in]\inf I(S), \sup I(Q(R))]$, and is a critical value of I_k.

3. EXISTENCE OF A CRITICAL POINT OF THE PENALIZED FUNCTIONAL

In this section we shall prove the existence of a critical point of the penalized functional, using Theorem 2.2. We shall show that the functional f_δ, $\delta > 0$, satisfies assumptions (i)-(iii) of Theorem 2.2.

In order to prove assumption (i), we need the following Lemma, whose proof is contained in [BFG3]

LEMMA 3.1 - *Let $\{t_m(s)\}$ be a sequence in $H^{1,2}(t_0, t_1;]a, b[)$, such that*

$$\sup_{m \in \mathbb{N}} \|\dot{t}_m\| < +\infty; \qquad (3.1)$$

There exists a sequence $(s_m)_{m \in \mathbb{N}}$ such that

$$\lim_{m \to \infty} \varphi(t_m(s_m)) = 0. \qquad (3.2)$$

Then:

$$\lim_{m \to \infty} \int_0^1 \frac{1}{\varphi^2(t_m)} ds = +\infty. \qquad (3.3)$$

Using Lemma 3.1, it is easy to prove the following

LEMMA 3.2 - *For every couple of real numbers* $a < b$, *the set*

$$f_a^b = \{z \in Z \mid a \leq f(z) \leq b\}$$

is a complete subset of Z.

In the following we shall assume for simplicity that

$$\langle \alpha(z)\xi, \xi \rangle = \alpha(z)\langle \xi, \xi \rangle,$$

i.e. α is a smooth function on M. The same calculation can be carried out also in the general case, with some more technicality.

In the next Lemma we show that for every $\delta > 0$, the penalized functional f_δ satisfies (P.S.) condition.

LEMMA 3.3 - *For every* $\delta > 0$, *the penalized functional* f_δ *satisfies* (P.S.).

PROOF. - Let $(z_m)_{m \in \mathbb{N}} = (x_m, t_m)_{m \in \mathbb{N}}$ be a sequence in Z, such that

$$|f(z_m)| \leq c, \qquad \forall m \in \mathbb{N}, \qquad (3.4)$$

$$\|f'(z_m)\|_{Z_m} \xrightarrow[m \to \infty]{} 0. \qquad (3.5)$$

For every $m \in \mathbb{N}$, we set

$$\tau_m = t_m - \bar{t},$$

where $\bar{t}$ is the segment joining t_0 and t_1. Denoting by $o(1)$ an infinitesimal sequence, A_4) and (3.5) give:

$$o(1)\|\tau_m\|_1 = f'_\delta(z_m)[(0, \tau_m)] =$$

$$= \frac{1}{2}\int_0^1 \alpha_t(z_m)\langle \dot{x}_m, \dot{x}_m\rangle \tau_m ds - \frac{1}{2}\int_0^1 \beta_t(z_m)\dot{t}_m^2 \tau_m ds +$$

$$- \int_0^1 \beta(z_m)\dot{t}_m\dot{\tau}_m ds - \delta\|\dot{t}_m\|^2 \int_0^1 \dot{t}_m \dot{\tau}_m ds +$$

$$+ 2\delta \int_0^1 \frac{\varphi'(t_m)}{\varphi^3(t_m)} \tau_m ds \le$$

$$\le L\|\tau_m\|_\infty \left[\|\dot{x}_m\|^2 + \|\dot{t}_m\|^2\right] - \int_0^1 \beta(z_m)\dot{t}_m^2 ds + \int_0^1 \beta(z_m)\dot{t}_m \bar{t} ds +$$

$$- \delta\|\dot{t}_m\|^2 \int_0^1 \dot{t}_m^2 ds + \delta\|\dot{t}_m\|^2 \int_0^1 \dot{t}_m \bar{t} ds + 2\delta \int_0^1 \frac{\varphi'(t_m)}{\varphi^3(t_m)} (t_m - \bar{t}) ds,$$

where $\|\tau_m\|_\infty = \sup |\tau_m(s)|$.

Then, putting $\Delta = t_1 - t_0$, A_3) gives

$$o(1)\|\tau_m\|_1 \le$$

$$\le L\|\tau_m\|_\infty \left[\|\dot{x}_m\|^2 + \|\dot{t}_m\|^2\right] - \int_0^1 \beta(z_m)\dot{t}_m^2 ds + \Delta\|\dot{t}_m\| +$$

$$- \delta\|\dot{t}_m\|^4 + \delta\Delta^2\|\dot{t}_m\|^2 + 2\delta \int_0^1 \frac{\varphi'(t_m)}{\varphi^3(t_m)} (t_m - \bar{t}) ds. \tag{3.6}$$

Now, (3.4) gives

$$-\int_0^1 \beta(z_m)\dot{t}_m^2 \le -\frac{\delta}{2}\left[\|\dot{x}_m\|^4 - \|\dot{t}_m\|^4\right] + 2\delta \int_0^1 \frac{1}{\varphi^2(t_m)} ds + c \tag{3.7}$$

and (3.6) and (3.7) give:

$$o(1)\|\tau_m\|_1 \le$$

$$L\|\tau_m\|_\infty \left[\|\dot{x}_m\|^2 + \|\dot{t}_m\|^2\right] + \Delta\|\dot{t}_m\| + \delta\Delta^2\|\dot{t}_m\|^2 - \frac{\delta}{2}\left[\|\dot{x}_m\|^4 + \|\dot{t}_m\|^4\right] +$$

$$+ 2\delta \int_0^1 \frac{1}{\varphi^2(t_m)} ds + 2\delta \int_0^1 \frac{\varphi'(t_m)}{\varphi^3(t_m)} (t_m - \bar{t}) ds + c. \tag{3.8}$$

Now, by the choice of the function $\varphi(t) = (t-a)(b-t)$ and of $\bar{t}$, it can be proved that there exists a constant c_1 such that

$$2\delta \int_0^1 \frac{1}{\varphi^2(t_m)} ds + 2\delta \int_0^1 \frac{\varphi'(t_m)}{\varphi^3(t_m)}(t_m - \bar{t}) ds \le c_1, \qquad (3.9)$$

hence (3.8) and (3.9) give:

$$\frac{\delta}{2}\left[\|\dot{x}_m\|^4 + \|\dot{t}_m\|^4\right] \le L\|\tau_m\|_\infty \left[\|\dot{x}_m\|^2 + \|\dot{t}_m\|^2\right] +$$

$$\Delta \|\dot{t}_m\| + \delta\Delta^2 \|\dot{t}_m\|^2 - o(1)\|\tau_m\|_1 + c + c_1,$$

from which we deduce that the sequences $(\|\dot{x}_m\|)_{m \in \mathbb{N}}$ and $(\|\dot{t}_m\|)_{m \in \mathbb{N}}$ are bounded. Moreover, from $A_2)$-$A_3)$ and (3.4),

$$\sup_{m \in \mathbb{N}} \int_0^1 \frac{1}{\varphi^2(t_m)} ds < +\infty. \qquad (3.10)$$

From the boundness of $(\|\dot{x}_m\|)_{m \in \mathbb{N}}$ and $(\|\dot{t}_m\|)_{m \in \mathbb{N}}$, we have that the sequences $(x_m)_{m \in \mathbb{N}}$ and $(t_m)_{m \in \mathbb{N}}$ weakly converge to $x \in H^{1,2}(I, \mathbb{R}^N)$ and $t \in H^{1,2}(I, \mathbb{R})$ respectively. By $A_1)$ we have that $x \in \Omega^1$. Morever by (3.10) and Lemma 3.2, $t \in H^{1,2}(t_0, t_1;]a, b[)$. Finally, using the arguments used for proving Proposition 3.3 of [BFM2], the sequences $(x_m)_{m \in \mathbb{N}}$ and $(t_m)_{m \in \mathbb{N}}$ strongly converge to $x \in H^{1,2}(I, \mathbb{R}^N)$ and $t \in H^{1,2}(I, \mathbb{R})$ respectively, so the proof of the Lemma is complete. ∎

REMARK 3.4 - With the same proof of Lemma 3.3, for every $k \in \mathbb{N}$, the restriction $f_{\delta,k}$ of f_δ to Z_k satisfies (P.S.).

LEMMA 3.5 - With the notations of section 2 (see (2.7), (2.8), (2.10)), the unpenalized action integral f verifies the following properties:

a) $\qquad\qquad\qquad \inf f(S) > -\infty;$

b) $\qquad\qquad\qquad \sup f(Q(R)) < +\infty, \forall R > 0;$

c) $\exists \overline{R} > 0$, such that

$$\sup f(\partial Q(\overline{R})) < \inf f(S).$$

PROOF. - a). By virtue of A_3), for every $z = (x,\overline{t}) \in S$, we have:

$$f(z) = \frac{1}{2}\int_0^1 \left[\alpha(z)\langle \dot{x},\dot{x}\rangle - \beta(z)\dot{t}^2\right]ds \geq -\Delta^2,$$

where $\Delta = t_1 - t_0$.

b). By assumption A_4), there exists a continous function $a(x)$ such that for every $z = (x,t) \in M$:

$$\alpha(x,t) \leq a(x)$$

Hence, for every $z = (\overline{x},t) \in Q(R)$, we get:

$$f(z) \leq c_1 \int_0^1 a(\overline{x})ds - \nu \int_0^1 \dot{t}^2 ds, \qquad (3.11)$$

where c_1 is a suitable constant. From (5.2) we deduce b).

c). From (5.2) we have for every $z \in \partial(Q(R))$:

$$f(z) \leq c_2 - \nu\|\dot{t}\|^2 \longrightarrow -\infty \text{ as } \|\dot{t}\| \longrightarrow +\infty. \qquad (3.12)$$

Hence, from point a) and (3.12), we get c) for $\overline{R}$ sufficiently large. ●

Similar calculations show that this lemma holds for the penalized functionals f_δ, too. Next lemma gives more precise results if δ is sufficiently small.

LEMMA 3.6 - $\exists \delta_0 > 0$, such that $\forall \delta < \delta_0$:

$$\sup f_\delta(\partial Q(\overline{R})) < \inf f_\delta(S),$$

where $\overline{R}$ is the constant defined in c) of Lemma 3.5 and is independent on $\delta \in [0,\delta_0]$.

PROOF. - Let

$$l = \sup f(\partial Q(\overline{R})) < \inf f(S) = l'.$$

For every $(x,\overline{t}) \in S$ and $(\overline{x},t) \in \partial Q(\overline{R})$:

$$f_\delta(\overline{x},t) - f_\delta(x,\overline{t}) = f(\overline{x},t) - f(x,\overline{t}) +$$

$$+ \frac{\delta}{4}\left[\|\dot{x}\|^4 + \|\dot{t}\|^4\right] + \delta\int_0^1 \frac{1}{\varphi^2(t)}ds - \delta\left[\|\dot{\overline{x}}\|^4 + \|\dot{\overline{t}}\|^4 + \int_0^1 \frac{1}{\varphi^2(\overline{t})}ds\right] \geq$$

$$\geq l' - l - \delta c_1,$$

where c_1 is a constant independent on δ. Hence if δ is small enough, for every $(x,\overline{t}) \in S$ and $(\overline{x},t) \in \partial Q(\overline{R})$, we get:

$$f_\delta(\overline{x},t) - f_\delta(x,\overline{t}) \geq \frac{l'-l}{2},$$

and the Lemma is proved. ●

Combining Lemma 3.2, Lemma 3.3 and Lemma 3.6, Theorem 2.2 gives the following

THEOREM 3.7 - *For every $\delta \in \,]0,\delta_0[$ and $k \in \mathbb{N}$, the restriction $f_{\delta,k}$ of the penalized functional f_δ to the submanifold Z_k (see (2.6)), has a critical point $z_{\delta,k}$, such that:*

$$f_{\delta,k}(z_{\delta,k}) \leq \sup f_\delta(Q(\overline{R})) \leq M, \qquad (3.13)$$

where M is a suitable constant independent on $\delta \in \,]0,\delta_0]$.

PROOF. - The existence of a critical point $z_{\delta,k}$ such that $f_{\delta,k}(z_{\delta,k}) \leq \sup f_\delta(Q(\overline{R}))$ is consequence of Theorem 2.2 and Lemma 3.6. Moreover, for every $\delta \in \,]0,\delta_0]$,

$$\sup f_\delta(Q(\overline{R})) \leq \sup f(Q(\overline{R})) + \delta_0 \|\dot{\overline{x}}\|^4 = M,$$

and (3.13) is proved. ●

Now, using (3.13) and the arguments used for proving Lemma 3.3, it is possible to get the limit of the sequence $(z_{\delta,k})_{k\in\mathbb{N}}$. Moreover, using the arguments of Theorem 5.5 of [BFM2], the limit is a critical point of the penalized functional f_δ. The following Theorem holds:

THEOREM 3.8 - *For every $\delta \in]0, \delta_0]$, the penalized functional f_δ has a critical point z_δ such that*

$$f_\delta(z_\delta) \le M, \tag{3.14}$$

where M is a constant independent on δ (see (3.13)).

4. SOME ESTIMATES ON THE CRITICAL POINTS OF THE PENALIZED FUNCTIONAL

In this section some estimates on the critical points of the penalized functionals are proved. Assumption A_5) play a basic role in order to get these estimates, in particular in the following

THEOREM 4.1 - *Let $z = (x,t)$ be a critical point of the functional f_δ, and let η be the positive number defined in A_5) (η can be chosen small enough, in particular $\eta < \{|t_0|, |t_1|, \frac{b-a}{2}\}$). Then, for every $s \in I$:*

$$t(s) - a \ge \eta, \tag{4.1}$$

$$b - t(s) \ge \eta. \tag{4.2}$$

PROOF. - Since z is a critical point of f_δ, it is a smooth curve. Moreover, it is easy to see that t satisfies the following differential equation:

$$\frac{d}{ds}[\beta(z)\dot{t} + \delta\|\dot{t}\|^2\dot{t}] + \frac{1}{2}\langle\dot{x},\dot{x}\rangle\alpha_t(z) +$$

$$- \frac{1}{2}\dot{t}^2\beta_t(z) + 2\delta\frac{\varphi'(t)}{\varphi^3(t)} = 0. \tag{4.3}$$

Now, let $\rho(s) = \varphi(t(s))$, and let s_0 be a minimum point for ρ. Then the thorem is proved if (4.1) and (4.2) hold for $s = s_0$. If this is not true, and suppose for instance that we have:

$$0 \le \rho''(s_0) = \varphi''(t(s_0))\dot{t}^2(s_0)) + \varphi'(t(s_0))\ddot{t}(s_0). \qquad (4.5)$$

Since $\dot{t}(s_0) = 0$, (4.3) and (4.5) give:

$$0 \le \varphi'(t(s_0))\ddot{t}(s_0)) =$$

$$= -\frac{\varphi'(t(s_0))}{\beta(z(s_0)) + \delta\|\dot{t}\|^2}\left[\frac{1}{2}\langle \dot{x}(s_0),\dot{x}(s_0)\rangle \alpha_t(z(s_0)) + 2\delta\frac{\varphi'(t(s_0))}{\varphi^3(t(s_0))}\right].$$

On the other hand, ii) of A_5), (4.4), and the definition of φ give

$$\varphi'(t(s_0))\alpha_t(z(s_0)) \ge 0,$$

hence

$$0 \le -\frac{\varphi'(t(s_0))}{\beta(z(s_0)) + \delta\|\dot{t}\|^2} \cdot \frac{\varphi'(t(s_0))}{\varphi^3(t(s_0))} < 0,$$

which is a contradiction. In the same way, a contradiction is obtained assuming that

$$b - t(s_0) < \eta. \blacksquare$$

REMARK 4.2 - Lemma 4.1 says that all the critical points of the penalized functional f_δ are uniformly far from the boundary $\partial M = M_0 \times \{a\} \cup M_0 \times \{a\}$. Assumption A_5) is a slight stronger condition than the convexity property of the boundary of M (see [Ma] for the convexity of a strip).

REMARK 4.3 - Notice that Theorem 4.1 implies that the quantities

$$\frac{1}{\varphi^2(t)}, \frac{\varphi'(t)}{\varphi^3(t)},$$

are uniformly bounded with respect to the critical point $z = (x,t)$ of f_δ and $\delta > 0$.

Consider now a critical point $z = (x,t)$ of f_δ. Then, like a critical point of the action integral (i.e. a geodesic), a conservation law is satisfied by z. Indeed the following Lemma, whose proof is consequence of the criticality of z, holds (see [BFM2]).

LEMMA 4.4 - *There exists a constant $E_\delta(z)$, such that for every $s \in I$:*

$$E_\delta(z) = \alpha(z(s)))\langle\dot{x}(s),\dot{x}(s)\rangle - \beta(z(s))\dot{t}(s)^2 +$$

$$+ \delta[\|x\|^2\langle\dot{x}(s),\dot{x}(s)\rangle - \|t\|^2\dot{t}(s)^2] + 2\delta\frac{1}{\varphi^2(t(s))}.$$

Consider now for a critical point $z = (x,t)$ of f_δ the following quantity:

$$\tilde{E}_\delta(z) = E_\delta(z) - \delta\|x\|^2\langle\dot{x}(s),\dot{x}(s)\rangle. \qquad (4.6)$$

Notice that $\tilde{E}_\delta(z)$ is not conserved along z. However it is useful to prove the following estimate on the derivative of z.

LEMMA 4.5 - *Let $c \in \mathbb{R}$ and $\delta_0 > 0$, then there exists a constant $K = K(c, \delta_0) > 0$, such that for every critical point $z = (x,t)$ of f_δ, $\delta \in [0,\delta_0]$, with $\tilde{E}_\delta(z) \le c$:*

$$\|\dot{t}\|_\infty \le K. \qquad (4.7)$$

Moreover, there exists a constant $K' = K'(c,\delta_0)$, such that

$$\|\dot{x}\|_\infty \le K'. \qquad (4.8)$$

PROOF. - Put

$$u(s) = \beta(z)\dot{t} + \delta\|t\|^2\dot{t}, \qquad (4.9)$$

then (4.3) gives:

$$u'(s) = -\frac{1}{2}\langle\dot{x},\dot{x}\rangle\alpha_t(z) + \frac{1}{2}\dot{t}^2\beta_t(z) - 2\delta\frac{\varphi'(t)}{\varphi^3(t)}. \qquad (4.10)$$

Moreover, by $A_2)-A_4)$, there exists a constant $\Lambda > 0$, such that for every $z \in M$:

$$\frac{|\beta_t(z)|}{\beta(z)} \leq \Lambda, \qquad (4.11)$$

$$\frac{|\alpha_t(z)|}{\alpha(z)} \leq \Lambda. \qquad (4.12)$$

We set for simplicity $\Lambda = 1$.

Now, (4.10), (4.11) and (4.12) give

$$u'(s) \leq \alpha(z)\langle \dot{x},\dot{x}\rangle + \beta(z)\dot{t}^2 - 2\delta\frac{\varphi'(t)}{\varphi^3(t)}.$$

Moreover, by Remark 4.3 and (4.6), we have:

$$u'(s) \leq$$

$$\leq c + \beta(z(s))\dot{t}(s)^2 + \delta\|\dot{t}\|^2\dot{t}(s)^2 - 2\delta\frac{1}{\varphi^2(t(s))} + \beta(z)\dot{t}^2 - 2\delta\frac{\varphi'(t)}{\varphi^3(t)} \leq$$

$$\leq 2u(s)\dot{t}(s) + |c| + c',$$

where c' is a constant due to Lemma 4.1.

Now, using an argument similar to Gronwall Lemma (see also the proof of Theorem 4.1 of [BFM2]), we get (4.7). Moreover (4.6) and (4.7) easily give (4.8). ●

5. PROOF OF THEOREM 1.4

In this section we shall prove Theorem 1.4, obtaining the geodesic connectedness of a strip, under assumption $A_1)-A_5)$.

In section 3, it has been proved the existence of a critical point $z_\delta = (x_\delta, t_\delta)$, $\delta \in]0, \delta_0[$, of the penalized functional f_δ such that (3.14) holds. We shall show that this family contains a sequence which converges to a critical point of the action integral f, i.e. a geodesic joining z_0 and z_1. We need the following

GEODESIC CONNECTEDNESS 241

LEMMA 5.1 - Let $z_\delta = (x_\delta, t_\delta)$, $\delta \in \,]0, \delta_0[$, be a critical point of f_δ such that (3.14) holds. Then, there exists a constant $c \in \mathbb{R}$, such that for every $\delta \in \,]0, \delta_0[$:

$$\tilde{E}_\delta(z_\delta) \leq c.$$

PROOF. - Integrating (4.6) gives:

$$\tilde{E}_\delta(z_\delta) = \int_0^1 [\alpha(z_\delta(s))\langle \dot{x}_\delta(s), \dot{x}_\delta(s)\rangle - \beta(z_\delta(s))\dot{t}_\delta(s)^2] ds +$$

$$- \|\dot{t}_\delta\|^4 + 2\delta \int_0^1 \frac{1}{\varphi^2(t_\delta(s))} ds =$$

$$= 2f_\delta(z_\delta) + -\frac{\delta}{2}\|\dot{t}_\delta\|^2 + 4\delta \int_0^1 \frac{1}{\varphi^2(t_\delta(s))} ds \leq$$

$$2M + 4\delta \int_0^1 \frac{1}{\varphi^2(t_\delta(s))} ds.$$

By Remark 4.3, the integral $\int_0^1 \frac{1}{\varphi^2(t_\delta(s))} ds$ is uniformly bounded with respect to δ, and the lemma is proved. ●

PROOF OF THEOREM 1.4 - By Theorem 3.8, there exists a critical point $z_\delta = (x_\delta, t_\delta)$, $\delta \in \,]0, \delta_0[$, of f_δ, such that (3.14) holds. Moreover, by Lemma 5.1 and Lemma 4.5, there exist two constants K and K', such that (4.7) and (4.8) hold. Hence the families (x_δ) and (t_δ), $\delta \in \,]0, \delta_0[$, are bounded respectively in $H^{1,2}(I, \mathbb{R}^N)$ and $H^{1,2}(I, \mathbb{R})$, so it can be found a sequence $\delta_m \longrightarrow 0$, such that $(x_{\delta_m})_{m \in \mathbb{N}}$ and $(t_{\delta_m})_{m \in \mathbb{N}}$ weakly converge respectively to $x \in H^{1,2}(I, \mathbb{R}^N)$ and $t \in H^{1,2}(I, \mathbb{R})$. Since Ω^1 is a complete submanifold of $H^{1,2}(I, \mathbb{R}^N)$, $x \in \Omega^1$. Moreover, by Lemma 4.1, $t \in H^{1,2}(t_0, t_1;]a, b[)$. Hence $z = (x, t) \in Z$.

Using the arguments of Lemma 3.3, we have that the convergence of the sequences $(x_{\delta_m})_{m \in \mathbb{N}}$ and $(t_{\delta_m})_{m \in \mathbb{N}}$ is strong. From this fact, it follows that $z = (x, t)$ is a critical point of f.

Indeed, let $\zeta = (\xi,\tau) \in T_z Z$ and let ξ_m be the orthogonal projection of ξ on $T_{x_{\delta_m}}\Omega^1$. Then $(\xi_m)_{m \in \mathbb{N}}$ weakly converges to ξ in $H^{1,2}(I,\mathbb{R}^N)$ (see [BF2]). Since z_{δ_m} is a critical point of f_{δ_m}, we have:

$$0 = f'_{\delta_m}(z_{\delta_m})[\xi_m,\tau] =$$

$$\int_0^1 \left[\alpha(z_{\delta_m})<\dot{x}_{\delta_m},\dot{\xi}_m> + \frac{1}{2}<\dot{x}_{\delta_m},\dot{x}_{\delta_m}><\nabla\alpha(z_{\delta_m}),\xi_m> + \frac{1}{2}<\dot{x}_{\delta_m},\dot{x}_{\delta_m}>\alpha_t(z_{\delta_m})\tau \right] ds +$$

$$- \int_0^1 \left[\beta(z_{\delta_m})\dot{t}_{\delta_m}\dot{\tau} + \frac{1}{2}\dot{t}_{\delta_m}^2<\nabla\beta(z_{\delta_m}),\xi_m> + \frac{1}{2}\dot{t}_{\delta_m}^2\beta_t(z_{\delta_m})\tau \right] ds +$$

$$+ \delta_m \|\dot{x}_{\delta_m}\|^2 \int_0^1 <\dot{x}_{\delta_m},\dot{\xi}_m>ds - \delta_m\|\dot{t}_{\delta_m}\|^2 \int_0^1 \dot{t}_{\delta_m}\dot{\tau}ds + 2\delta_m \int_0^1 \varphi'(t_{\delta_m})\frac{\tau}{\varphi^3(t_{\delta_m})}ds,$$

where $\nabla\alpha(z)$ and $\nabla\beta(z)$ denote respectively the gradient of α and β with respect to the Riemannian structure of M_0.

Taking the limit above, we obtain

$$f'(z)[\zeta] = 0,$$

and the proof of Theorem 1.1 is complete. ∎

REFERENCES

[BP]   J.K.BEEM, R.E.PARKER, Pseudoconvexity and Geodesic connectedness, *Ann. Mat. Pura e Appl.* (IV), **CLV** 137-142 (1989).

[BF1] V.BENCI, D.FORTUNATO, Existence of geodesics for the Lorentz metric of a stationary gravitational field, *Ann. Inst. H.Poincare', Analyse non Lineaire* 7, (1990) 27-35.

[BF2] V.BENCI, D.FORTUNATO, On the existence of infinitely many geodesics on space-time manifolds, to appear on *Adv. Math.*

[BFG1] V.BENCI, D.FORTUNATO, F.GIANNONI, On the existence of multiple geodesics in static space-times, *Ann. Inst. H.*

Poincarè Analyse non lineaire **8**, 79-102 (1991).

[BFG2] V.BENCI,D.FORTUNATO,F.GIANNONI, Geodesic on static Lorentz manifolds with convex boundary, *Variational methods in Hamiltonian systems and elliptic equations*, M.Girardi, M.Matzeu Ed. (Pitman research notes in Mathematics), 21-41 (1992).

[BFG3] V.BENCI,D.FORTUNATO,F.GIANNONI, On the existence of geodesics in static Lorentz manifolds with nonsmooth boundary, *Ann. Sc. Norm. Sup. Pisa,* in press.

[BFM1] V.BENCI,D.FORTUNATO,A.MASIELLO, Geodesics in Lorentzian manifolds, preprint Dip. Mat. Univ. Bari, 5/92.

[BFM2] V.BENCI,D.FORTUNATO,A.MASIELLO, On the geodesic connectedeness of Lorentzian manifolds, preprint Dip. Mat. Univ. Bari, 9/92.

[Ge] R.GEROCH, Domains of dependence, *J. Math. Phys.* **11**, 437-449 (1970).

[Gi] F.GIANNONI, Geodesics on non static Lorentz manifolds of Reissner-Nordström type, *Math. Ann.* **291**, 383-401 (1991).

[GM] F.GIANNONI,A.MASIELLO, On the existence of geodesics on stationary Lorentz manifolds with convex boundary, *J. Funct. Anal.* **101**, 340-369 (1991).

[LL] L.LANDAU,E.LIFCHITZ, *Theorie des champs*, Mir, Moscou, 1970.

[Ma] A.MASIELLO, Convex regions in Lorentzian manifolds, in preparation.

[Na] J.NASH, The embedding problem for Riemannian manifolds, *Ann. of Math.* **63**, 20-63 (1956).

[ON] B.O'NEILL, *Semi-Riemannian geometry with applications to Relativity*, Academic Press Inc., New York-London, 1983.

[Pa] R.S.PALAIS, Morse Theory on Hilbert manifolds, *Topology*, **2**, 299-340 (1963).

[Pe] R.PENROSE, *Techniques of differential topology in relativity*, Conf. Board Math.Sci. **7**, S.I.A.M. Philadelphia, 1972.

[Ra]   P.H.RABINOWITZ, *MinMax methods in critical point theory with applications to Differential Equations*, CBMS Reg. Conf. Soc. in MATH. n. 65, AMS, 1984.

[U]   K.UHLENBECK, *A Morse theory for geodesics on a Lorentz manifold*, *Topology* **14**, (1975), 69-90.

MULTIPLE PERIODIC SOLUTIONS TO SOME N-BODY TYPE PROBLEMS
VIA A COLLISION INDEX

PIETRO MAJER* AND SUSANNA TERRACINI **

* SISSA, Strada Costiera 11, Trieste
** Dipartimento di Matematica del Politecnico, Piazza Leonardo da Vinci 32, Milano

ABSTRACT. In this paper we prove the existence of infinitely many periodic solutions to a class of "n-body-type" dynamical systems of the form:

$$-m_i \ddot{u}_i = \sum_{j=1, j\neq i}^{n} \nabla V_{ij}(u_i - u_j, t) \qquad (HS)$$

when the interaction potentials V_{ij} present at the origin an attractive singularity having a strong-force growth. We follow a variational approach, looking for periodic solutions of (HS) as critical points of the associated action integral. More precisely we prove the existence of an unbounded sequence of critical levels for the action integral. The proof is based on a variant of Ljusternik-Schnirelman variational theory. Some topological aspects of the loop space of the configuration space are investigated. To this end we introduce the notion of "collision index" and we prove the existence of sets having arbitrarily large collision index.

§1. INTRODUCTION AN STATEMENT OF THE RESULTS

In this paper we are concerned with periodic solutions to singular dynamical systems like (HS). From our point of view, such systems belong to the larger class of Hamiltonian Systems of the form

$$-Au = \nabla V(u, t) \qquad (1.1)$$

where $u = (u_1, \ldots, u_n) \in \mathbf{R}^{kn}$, $n \geq 2$, $k \geq 3$, A induces the quadratic form

$$\langle Au, u \rangle = \int_0^T \sum_{i=1}^n m_i |\dot{u}_i|^2$$

and V is the interaction potential, T-periodic in time,

$$V(x_1, \ldots, x_n, t) = \frac{1}{2} \sum_{i,j=1; i \neq j}^n V_{ij}(x_i - x_j, t),$$ (1.2)

$(x_1, \ldots, x_n) \in \mathbf{R}^{kn}, x_i \neq x_j, \forall i \neq j$.

REMARK: With no loss of generality we can assume that in (1.2) the V_{ij}'s verify $V_{ij}(x, t) = V_{ji}(-x, t)$ (Newton's III law of Mechanics). Then system (1.1) can be rewritten in the more familiar form (HS).

We shall assume that V is of class C^1 on its domain; thus T-periodic solutions of (1.1) are critical points of the associated action integral:

$$f(u) = \frac{1}{2} \langle Au, u \rangle - \int_0^T V(u, t)$$

whose natural domain is the function set

$$\Lambda = \{u = (u_1, \ldots, u_n) \in H^1_{T-per}(\mathbf{R}; \mathbf{R}^{kn}) : u_i(t) \neq u_j(t), \forall t, \forall i \neq j\}.$$

Our assumptions on the potentials V_{ij} are the following (for $i \neq j$):

$$V_{ij}(x, t) \leq 0 \qquad \forall x \in \mathbf{R}^k \setminus \{0\} \qquad (V1)$$

$$\exists \rho_0 > 0 \, \exists U \in C^1(\mathbf{R}^k \setminus \{0\}; \mathbf{R}), \lim_{x \to 0} U(x) = +\infty :$$
$$- V_{i,j}(x, t) \geq |\nabla U(x)|^2 + U(x), \qquad \forall x, 0 < |x| < \rho_0 \qquad (V2)$$

$$\exists \rho > 0 \, \exists \theta, 0 \leq \theta < \frac{\pi}{2} :$$
$$\text{ang}(\nabla V_{ij}(x, t), x) \leq \theta \qquad \forall x, |x| > \rho \qquad (V3)$$

where $\nabla V_{ij}(x, t)$ denotes $(\frac{\partial}{\partial x_1} V_{ij}(x, t), \ldots, \frac{\partial}{\partial x_n} V_{ij}(x, t)) \in \mathbf{R}^k$, and in any euclidean space $0 \leq \text{ang}(x, y) \leq \pi$ denotes the angle between x and y

Our main result is the following theorem:

THEOREM 1. *Assume (V1), (V2) and (V3) hold. Then the associated functional f has an unbounded sequence of critical levels. Therefore (1.1) possesses infinitely many periodic solutions in Λ.*

Assumptions (V1) and (V3) are quite natural when thinking to a model problem where $V_{ij}(x,t) = -a/|x|^\alpha$, for some positive α. Concerning condition (V2), -the "strong force" condition- it is introduced in order to obtain closed sublevels for the functional f. It is a quite unnatural hypotheses in the contests of Celestial Mechanics, since it implies a growth at the origin of order $-\alpha$, $\alpha \geq 2$ for the potentials. However, the relaxation of the strong force condition leads to very hard problems in the variational approach even to the two-body problem (see [7]) when the existence of noncollision solutions (i.e. belonging to Λ) is investigated. We wish to point out that systems like (1.1) are studied here from the point of view of Nonlinear Analysis rather than Celestial Mechanics; the interest of our results just rely on the weakness of the assumptions on the potentials (only growth conditions at the origin and the requirement on the direction of the force field at infinity). An analogous perspective has been adopted in several papers, mainly concerning the existence of periodic solutions to some two-body like problems: [1], [2], [4], [7], [8], [11], [12], [13]. Concerning the n-body problem, symmetrical cases (i.e. $V_{ij}(-x) = V_{ij}(x)$) have been studied in [3], [6], [8], [19]. A symmetry constraint on the function space allows to overcome the lack of compactness and the existence of one solution can be derived from a minimization argument (or a mountain-pass argument). The common problem of these papers is how to avoid collision solutions, when the strong force condition (V2) is weakened.

In the non symmetrical case, the three-body problem ($n = 3$) has been studied by Bahri and Rabinowitz in [5], where they prove the existence of infinitely many solutions under assumptions slightly different to the ones here. With some more computations (see the truncation argument in [16]) one can extend Theorem 1 to cover Bahri and Rabinowitz's case when $n \geq 4$.

We shall follow here the approach introduced in [16] where the existence of at least one T-periodic solution of (1.1) has been proved, under assumptions (V1)-(V3). We shall use a variant of the classical deformation lemma which permits to handle functionals f which do not satisfy themselves the usual Palais-Smale condition, while the Palais-Smale condition is fulfilled by their restrictions to some sublevel of an auxiliary functional g, provided a compatibility condition between ∇f and ∇g is satisfied. This deformation lemma provides critical points for f if, roughly, some sublevel of f is not deformable into a superlevel of g in itself. At this point, a description of some topological properties of the space Λ is needed. We point out that the usual Ljusternik-Schnirelman category is not the most suitable tool to describe the topological features of the loop space Λ. The reason is that every "neighborhood of infinity" (i.e., the complement of any set where the Palais-Smale condition holds) has infinite category. To overcome this we introduce a topological index ("collision index"), closely related to the relative category in the sense of [10], which seems to be more natural in the n-body contest. Finally we shall prove the existence of subsets having arbitrarily large collision index.

NOTATIONS:

n: number of the bodies $(u_1, \ldots, u_n)$
k: dimension of the space of each body: $u_i \in \mathbf{R}^k$, $k \geq 3$;
T: period.
H denotes the Sobolev space of periodic functions:

$$H = \{u(t) = (u_1(t), \ldots, u_n(t)) \in H^1_{loc}(\mathbf{R}; \mathbf{R}^{kn}) \; : \; u(t+T) = u(t), \forall t\}$$

endowed with the inner product

$$u \cdot v = \int_0^T (\dot{u}(t) \cdot \dot{v}(t) + u(t) \cdot v(t)) dt \,.$$

Λ is the subset of all the noncollision orbits:

$$\Lambda = \{u \in H \; : \; u_i(t) \neq u_j(t), \forall t, \forall i \neq j\} \,.$$

We identify the subspace of H of all the constant functions with $\mathbf{R}^{kn}$:

$$\mathbf{R}^{kn} \simeq \{u \in H \; : \; \dot{u}(t) = 0, \forall t\} \,.$$

For every function u, $[u]$ denotes its mean value:

$$[u] = \frac{1}{T} \int_0^T u(s) ds.$$

If $x, y \in \mathbf{R}^k$ the angle between x and y is

$$\text{ang}(x, y) = \begin{cases} \arccos \frac{x \cdot y}{|x||y|} & \text{if } |x||y| \neq 0 \\ 0 & \text{otherwise.} \end{cases}$$

For any functional $f : X \to \mathbf{R}$ we shall denote: $\{f \leq c\} = \{x \in X \; : \; f(x) \leq c\}$ and analogously for $\{f = c\}$ and $\{f \geq c\}$. K_c denotes the set of all the critical points of f at level c.

NOTA BENE: In what follows we shall assume for the sake of simplicity that $m_i = 1$. Simple variants on the estimates are needed to cover the general case.

§2 THE COLLISION INDEX

Let us fix some notations. $\mathcal{H}$ is the set of all the deformations of Λ in H into the subspace of constant functions $\mathbf{R}^{kn}$:

$$\mathcal{H} = \{h \in \mathcal{C}(\Lambda \times [0,1]; H) : h(\cdot,0) = \text{id}, h(\Lambda,1) \subset \mathbf{R}^{kn}\}. \tag{2.1}$$

To each pair (h,u) with $h = (h_1, \ldots, h_n) \in \mathcal{H}$ and $u = (u_1, \ldots, u_n) \in \Lambda$, we associate a relation $r_{h,u}$ in the set of all the indices, $I = \{1, \ldots, n\}$ in the following way:

$$i \, r_{h,u} \, j \iff \exists \lambda_{ij} \in [0,1] \, \exists t_{ij} \in S^1 : \\ h_i(u_1, \ldots, u_n, \lambda_{ij})(t_{ij}) = h_j(u_1, \ldots, u_n, \lambda_{ij})(t_{ij}). \tag{2.2}$$

It is immediate to check that $r_{h,u}$ is reflexive and symmetric. We denote by $R_{h,u}$ the smallest equivalence relation containing $r_{h,u}$: by definition we have that

$$i R_{h,u} j \iff \exists i_1, \ldots, i_k : i \, r_{h,u} \, i_1, i_1 \, r_{h,u} \, i_2, \ldots, i_k \, r_{h,u} \, j. \tag{2.3}$$

If the right-hand side of (2.2) holds true, we say that a collision between the ith body and the jth body occurs along the homotopy h. Thus two indices (i,j) are in relation R if they are connected by a chain of collisions.

The concept of *admissible set* was introduced in [16]:

DEFINITION 2.1: *Let A be a closed subset of Λ. We say that A is admissible if*

$$\forall h = (h_1, \ldots, h_n) \in \mathcal{H} \quad \exists u = (u_1, \ldots, u_n) \in A : \\ i \, R_{h,u} \, j, \, \forall i, j \in I. \tag{2.4}$$

REMARK: Of course we have

$$i \, r_{h,u} \, j \iff \min_{\substack{\lambda \in [0,1] \\ t \in [0,T]}} |h_i(u,\lambda)(t) - h_j(u,\lambda)(t)| = 0$$

therefore

$$i R_{h,u} j \iff \min_{\substack{(i_1,\ldots,i_n) \in I^n \\ i_1=i, i_n=j}} \max_{1 \le m \le n-1} \min_{\substack{\lambda \in [0,1] \\ t \in [0,T]}} |h_{i_{m+1}}(u,\lambda)(t) - h_{i_m}(u,\lambda)(t)| = 0$$

Let us define the map

$$\Phi(h,u) =$$
$$\max_{\substack{1 \leq i < j \leq n \\ i_1 = i, i_n = j}} \min_{(i_1,\ldots,i_n) \in I^n} \max_{1 \leq m \leq n-1} \min_{\substack{\lambda \in [0,1] \\ t \in [0,T]}} |h_{i_{m+1}}(u,\lambda)(t) - h_{i_m}(u,\lambda)(t)|. \quad (2.5)$$

Then

$$\Phi(h,u) = 0 \iff \forall i,j \in I \; i \, R_{h,u} \, j.$$

and

$$A \text{ is admissibile if and only if}$$
$$\forall h \in \mathcal{H} \, \exists u \in A \; : \; \Phi(h,u) = 0. \quad (2.6)$$

DEFINITION 2.2 *Let Σ be the class of all closed subsets of Λ. The collision index i is the map $i : \Sigma \to \mathbf{N} \cup \{+\infty\}$ defined as follows:*

(i) $i(A) = 0 \iff A = \emptyset$
(ii) $i(A) = +\infty \iff \forall \alpha \in \mathbf{N} \; \exists (h^1, \ldots, h^\alpha) \in (\mathcal{H})^\alpha :$
      $\forall u \in A \, \exists \ell \in \{1, \ldots, \alpha\}, R_{h^\ell, u}$ is the trivial relation
(iii) $i(A) = +\alpha \iff \alpha$ is the smallest integer $\exists (h^1, \ldots, h^\alpha) \in (\mathcal{H})^\alpha :$
      $\forall u \in A \, \exists \ell \in \{1, \ldots, \alpha\}, R_{h^\ell, u}$ is the trivial relation

REMARK 2.1. In other words $i(A) = \alpha$ if and only if α is the smallest integer such that there are $(h^1, \ldots, h^\alpha) \in (\mathcal{H})^\alpha$ such that

$$\forall u \in A \quad \max_{\ell \in \{1,\ldots,\alpha\}} \Phi(h^\ell, u) > 0. \quad (2.7)$$

REMARK 2.2. Of course a closed subset A is admissible if and only if $i(A) \geq 2$.

Here below we list the main properties of the collision index.

PROPOSITION 2.1. *Let $A, B \in \Sigma$: we have*

(i) $A \subseteq B \implies i(A) \leq i(B);$

(ii) if $h \in \mathcal{C}(\Lambda \times [0,1]; \Lambda)$ with $h(\cdot, 0) = \mathrm{id}$ then $i(\overline{h(A,1)}) \geq i(A);$

(iii) $i(A \cup B) \leq i(A) + i(B);$

(iv)   if $i(A) < +\infty$ then $i(A)$ is equal to the smallest integer α such that A is covered by the union of α closed non-admissible subset of Λ;

(v)   every $A \neq \emptyset$ possesses an open neighbourhood A' such that $i(A) = i(\overline{A'})$;

(vi)   if A is compact then there is $i(A) < +\infty$;

(vii) $i(A) \leq \text{cat}_\Lambda(A);$

(viii) $i(A) \leq \text{cat}_{\Lambda,\mathbf{R}^{kn}}(A) + 1.$

PROOF: Part (i) is an immediate consequence of the definition.
(ii) We assume that $i(\overline{h(A,1)}) = \alpha < +\infty$ (if not the proof is trivial). By definition there are $(h^1, \ldots, h^\alpha) \in (\mathcal{H})^\alpha$ such that, in particular,

$$\forall u \in A, \quad \max_{\ell \in \{1,\ldots,\alpha\}} \Phi(h^\ell, h(u,1)) > 0.$$

Define, for $\ell = 1, \ldots, \alpha$ the juxtaposition of h with h^ℓ:

$$\tilde{h}^\ell(u,\lambda) = \begin{cases} h(u,\lambda) & \text{if } \lambda \in [0, 1/2] \\ h^\ell(h(u,1), 2\lambda - 1) & \text{if } \lambda \in [1/2, 1]; \end{cases}$$

then $(\tilde{h}^1, \ldots, \tilde{h}^\alpha) \in (\mathcal{H})^\alpha$. Since h is an homotopy of Λ into Λ, there are not collisions for $0 \leq \lambda \leq 1/2$; we then deduce that

$$\forall u \in A, \quad \max_{\ell \in \{1,\ldots,\alpha\}} \Phi(\tilde{h}^\ell, u) > 0.$$

Thus $i(A) \leq \alpha$.
(iii) Let $i(A) = \alpha < +\infty$ and $i(B) = \beta < +\infty$ (if either $i(A) = +\infty$ or $i(B) = +\infty$ the claim is obviously true from (i)). Then there are $(h^1, \ldots, h^\alpha) \in (\mathcal{H})^\alpha$ and there are $(k^1, \ldots, k^\beta) \in (\mathcal{H})^\beta$ such that

$$\forall u \in A, \quad \max_{\ell \in \{1,\ldots,\alpha\}} \Phi(h^\ell, u) > 0$$

$$\forall u \in B, \quad \max_{\ell \in \{1,\ldots,\beta\}} \Phi(k^\ell, u) > 0.$$

Let us put $k^1 = h^{\alpha+1},\ldots, k^\beta = h^{\alpha+\beta}$; then we have found $\alpha + \beta$ homotopies in $\mathcal{H}$ such that

$$\forall u \in A \cup B \quad \max_{\ell \in \{1,\ldots,\alpha+\beta\}} \Phi(h^\ell, u) > 0.$$

Therefore $i(A \cup B) \leq \alpha + \beta$.

(iv) Assume first that $A \subseteq A_1 \cup \ldots \cup A_\alpha$ and that each A_ℓ is a closed non-admissible subset of Λ. Then, by definition, for every $\ell \in \{1,\ldots,\alpha\}$ there is $h^\ell \in \mathcal{H}$ such that $\forall u \in A_\ell$ $\Phi(h^\ell, u) > 0$. Thus we have that $\max_{\ell \in \{1,\ldots,\alpha\}} \Phi(h^\ell, u) > 0$, $\forall u \in A \subseteq A_1 \cup \ldots \cup A_\alpha$. Therefore $i(A) \leq \alpha$.

Now we prove that, whenever $i(A) = \alpha$, then A can be covered by α closed non-admissible subsets of Λ. Of course we can assume that $\alpha < +\infty$ and that there are $(h^1, \ldots, h^\alpha) \in (\mathcal{H})^\alpha$ such that

$$\forall u \in A, \quad \max_{\ell \in \{1,\ldots,\alpha\}} \Phi(h^\ell, u) > 0.$$

Now we observe that each map $\Phi(h^\ell, \cdot) : A \to \mathbf{R}$ is continuous. Thus, setting

$$A_\ell = \{u \in A : \Phi(h^\ell, u) > 0\},$$

the family $\{A_\ell\}$ is an open covering of A. As such it admits a refinement (but still covering) family $\{B_\ell\}$ such that each B_ℓ is closed in A (and thereby in Λ); indeed, as a topological subspace of Λ, A is a normal space and the closed refinement theorem applies. Of course each B_ℓ is a closed non-admissible set and $A = B_1 \cup \ldots \cup B_\alpha$.

(v) We prove that each non-admissible closed set $A \neq \emptyset$ possesses an open neighborhood A' such that $\overline{A'}$ is non-admissible. This will prove (v), taking into account part (iv). By (6) A is non-admissible if there is $h \in \mathcal{H}$ such that $\forall u \in A, \Phi(h, u) > 0$. Let us set $\Omega_0 = \{u \in \Lambda : \Phi(h, u) = 0\}$; since $\Phi(h, \cdot)$ is a continuous map, Ω_0 is closed. Thus, by well-known separation properties of metric spaces (as Λ is), A and Ω_0 can be separated, that is, there is an open neighborhood A' of A such that $\overline{A'} \cap \Omega_0 = \emptyset$. Thus $\overline{A'}$ is non-admissible too.

(vi) For every $x \in A$ there is $\varepsilon(x)$ such that the closed ball $\overline{B(x, \varepsilon(x))}$ is deformable into the point $\{x\}$ without intersecting $\partial \Lambda$. Since Λ is arcwise connected this means that each $\overline{B(x, \varepsilon(x))}$ is a non-admissible subset of Λ.

Now, if A is compact, the covering $(B(x, \varepsilon(x)))_{x \in A}$ possesses a finite subcovering $(B(x_i, \varepsilon(x)))_{i=1,\ldots,\alpha}$. Therefore $i(A) \leq \alpha$.

(vii) and (viii) are easy consequences of (iv), since if a closed set is deformable into a single point in Λ it is a fortiori non-admissible (remember that Λ is arcwise connected). ∎

§3. COLLISIONS AGAINST A FIXED BODY

DEFINITION 3.1. *Let $A \subset \Lambda$ be closed. We say that A is admissible with respect to the i_0th body ($i_0 \in \{1, \ldots, n\}$) if*

$$\forall h = (h_1, \ldots, h_n) \in \mathcal{H} \; \exists u \in A \, , \forall i \in \{1, \ldots, n\} \; i \neq i_0 \\ \exists \lambda_i \in [0, 1] \, , \exists t_i \in [0, T] \; : \; h_i(u, \lambda_i)(t_i) = h_{i_0}(u, \lambda_i)(t_i) \quad (3.1)$$

A short comparison between Definitions 2.1 and 3.1 is in order. In Definition 3.1 below we ask that, along any homotopy $h \in \mathcal{H}$, each body h_i has a collision with the i_0th body and we don't consider collisions against other bodies but that one. In Definition 2.1 we only require that the collision relation $r_{h,u}$ connect each pair of indices (i, j) with a chain of collisions. Hence the requirement of Definition 3.1 is strictly more restrictive that the one of Definition 2.1.

REMARK. For any $h \in \mathcal{H}$ and $u \in \Lambda$ let us define the continuous map

$$\Psi(i_0; h, u) = \max_{1 \leq i \leq n} \min_{\substack{\lambda \in [0,1] \\ t \in [0,T]}} |h_i(u, \lambda)(t) - h_{i_0}(u, \lambda)(t)| \, . \quad (3.2)$$

Then A is admissible with respect to the i_0th index if and only if

$$\forall h \in \mathcal{H} \; \exists u \in A \; : \; \Psi(i_0, h, u) = 0. \quad (3.3)$$

DEFINITION 3.2 *Let Σ be the class of all closed subsets of Λ and let $i_0 \in \{1, \ldots, n\}$ be fixed. The collision index with respect to the i_0th body $i(i_0; \cdot)$ is the map $i(i_0; \cdot) : \Sigma \to \mathbb{N} \cup \{+\infty\}$ defined as follows:*

(i) $i(i_0; A) = 0 \iff A = \emptyset$

(ii) $i(i_0; A) = +\infty \iff \forall \alpha \in \mathbb{N} \; \exists (h^1, \ldots, h^\alpha) \in (\mathcal{H})^\alpha$:
$\forall u \in A \; \exists \ell \in \{1, \ldots, \alpha\},$
$\exists \lambda_i^\ell \in [0, 1] \; \exists t_i^\ell \in [0, T] \; : \; h_i(u, \lambda_i^\ell)(t_i^\ell) = h_{i_0}(u, \lambda_i^\ell)(t_i^\ell)$

(iii) $i(i_0; A) = \alpha \iff \alpha$ is the smallest integer
$\exists (h^1, \ldots, h^\alpha) \in (\mathcal{H})^\alpha \; : \; \forall u \in A \; \exists \ell \in \{1, \ldots, \alpha\},$
$\exists i \neq i_0 \; \forall \lambda \in [0, 1] \; \forall t \in [0, T] \; : \; h_i(u, \lambda)(t) \neq h_{i_0}(u, \lambda)(t)$

REMARK 3.1. In other words $i(i_0; A) = \alpha$ if and only if α is the smallest integer such that there are $(h^1, \ldots, h^\alpha) \in (\mathcal{H})^\alpha$ such that

$$\forall u \in A \quad \max_{\ell \in \{1,\ldots,\alpha\}} \Psi(h^\ell, u) > 0. \qquad (3.4)$$

REMARK 3.2. Of course a closed subset A is admissible with respect to the i_0th body if and only if $i(i_0; A) \geq 2$.

REMARK 3.3. For every $A \in \Sigma$ we have $i(i_0; A) \geq i(A)$. Indeed if A is non-admissible it is non-admissible with respect to any fixed body.

The following properties can be proven arguing as in the proof of Proposition 2.1.

PROPOSITION 3.1. Let $A, B \in \Sigma$: we have

(i) $A \subseteq B \implies i(i_0; A) \leq i(i_0; B)$;

(ii) if $h \in \mathcal{C}(\Lambda \times [0,1]; \Lambda)$ with $h(\cdot, 0) = \mathrm{id}$ then $i(i_0; \overline{h(A,1)}) \geq i(i_0; A)$;

(iii) $i(i_0; A \cup B) \leq i(i_0; A) + i(i_0; B)$;

(iv) if $i(i_0; A) < +\infty$ then $i(i_0; A)$ is equal to the smallest integer α such that A is covered by the union of α closed non-i_0-admissible subset of Λ;

(v) every $A \neq \emptyset$ possesses an open neighbourhood A' such that $i(A) = i(i_0; \overline{A'})$;

(vi) if A is compact then there is $i(i_0; A) < +\infty$;

(vii) $i(i_0; A) \leq \mathrm{cat}_\Lambda(A)$;

(viii) $i(i_0; A) \leq \mathrm{cat}_{\Lambda, \mathbf{R}^{kn}}(A) + 1$.

The main goal of this section is the following theorem:

THEOREM 3.1. For every $\alpha \in \mathbf{N}$ and for every fixed body i_0, Λ contains at least one compact subset A such that $i_{i_0}(A) \geq \alpha$.

PROOF: We are going to use the following result, due to Fadell and Husseini [9]: let us denote by $\Lambda\left(S^{k-1}\right)$ the space of the free loop space on the sphere S^{k-1} (endowed with the uniform topology). Then, for every $\alpha \geq 1$ there exists a subset A of $\Lambda(S^{k-1})$ such that

$$\operatorname{cat}_{\Lambda(S^{k-1})}(A) = \alpha ; \qquad (3.5)$$

moreover, as it has already pointed out in [9], A can be took compact with respect to the $H^1(S^1; S^{k-1})$ topology. Now we consider the subset of H parametrized over A in the following way:

$$u_1(v)(t) = 0 \qquad \forall v \in A$$
$$u_i(v)(t) = (i-1)v(Tt) \qquad \forall i = 2, \ldots, n, \forall v \in A$$

We denote $A^* = \{(u = (u_1(v), \ldots, u_n(v)) : v \in A\}$. Of course $A^* \subset \Lambda$: we claim that

$$i_1(A^*) \geq \frac{\alpha}{2(n-1)} .$$

Indeed, suppose on the contrary that

$$i_1(A^*) = \beta < \frac{\alpha}{2(n-1)} . \qquad (3.6)$$

By Definition 3.1 then there are $h^1, \ldots, h^\beta \in \mathcal{H}$ such that

$$\forall v \in A \, u = (u_1(v), \ldots, u_n(v)) \, \exists \ell \in \{1, \ldots, \beta\} \, \exists i \in \{2, \ldots, n\}$$
$$\forall \lambda \in [0, 1] \, \forall t \in [0, T] : h_i(u, \lambda)(t) - \neq h_1(u, \lambda)(t) . \qquad (3.7)$$

Let us define, for $i = 2, \ldots, n$, $\ell = 1, \ldots, \beta$

$$\xi_i^\ell(v, \lambda)(t) =$$
$$\frac{1}{i-1} \left(h_i((u_1(v), \ldots, u_n(v)), \lambda)(t/T) - h_1((u_1(v), \ldots, u_n(v), \lambda)(t/T) \right) ; \qquad (3.8)$$

it is immediate to check that $\xi_i^\ell : A \times [0, 1] \to H^1(S^1; \mathbf{R}^k)$ are continuous maps, and that $\xi_i^\ell(\cdot, 0) = \mathrm{id}_{|A}$ and $\xi_i^\ell(A, 1) \subset \mathbf{R}^k$ (thought as the space of constant loops), because of the fact that each $h^\ell \in \mathcal{H}$. Now let

$$B_i^\ell = \{v \in A : \xi_i^\ell(v) \neq 0, \forall \lambda \in [0, 1], \forall t \in S^1\} .$$

Because of (3.7), the family $\{B_i^\ell\}_{i,\ell}$ is an open covering of A (a compact metric space); as such, it admits a closed refinement $\{A_i^\ell\}_{i,\ell}$ which is still a covering of A. Now each homotopy

$$\frac{\xi_i^\ell}{|\xi_i^\ell|} : A_i^\ell \to \Lambda(S^{k-1})$$

deforms each A_i^ℓ into the set of constant loops of S^{k-1} in $\Lambda(S^{k-1})$.

Therefore $\mathrm{cat}_{\Lambda(S^{k-1})}(A_i^\ell) \leq 2$, $\forall i = 2, \ldots, n$, $\forall \ell = 1 \ldots, \beta$. Since the family $\{A_i^\ell\}_{i,\ell}$ is a covering of A, the subadditivity property of the category yields

$$\mathrm{cat}_{\Lambda(S^{k-1})}(A) \leq 2(n-1)\beta$$

and hence, from (3.6),

$$\mathrm{cat}_{\Lambda(S^{k-1})}(A) < \alpha$$

in contradiction with (3.5).

By Remark 3.3, a straightforward consequence of the above theorem is the following result:

COROLLARY 3.1 *For every $\alpha \in \mathbf{N}$, Λ contains at least one compact subset A such that $i(A) \geq \alpha$.*

§4. THE DEFORMATION LEMMA

The following result is a variant of the classical deformation lemma

LEMMA 4.1. *Let Λ be an open subset of any Hilbert space H and let $f \in C^1(\Lambda; \mathbf{R})$, $g \in C^2(\Lambda; \mathbf{R})$. Assume that there are c and $b \in \mathbf{R}$ such that:*

$$\lim_{\nu \to +\infty} f(x_\nu) = +\infty \qquad \text{if } x_\nu \to x_0 \in \partial\Lambda \text{ and } g(x_\nu) \text{ is bounded} \qquad (H1)$$

$$\nabla g(x) \neq 0 \qquad \forall x : g(x) = b, f(x) = c \qquad (H2)$$

every sequence (x_ν) in Λ such that $f(x_\nu) \to c$, $\limsup_{\nu \to \infty} g(x_\nu) \leq b$ and $\nabla f(x_\nu) \to 0$ possesses a converging subsequence $\qquad (H3)$

every sequence (x_ν) in Λ such that $f(x_\nu) \to c$, $g(x_\nu) \to b$ and

$$\nabla f(x_\nu) - \lambda_\nu \nabla g(x_\nu) \to 0,$$
with $\lambda_\nu \geq 0$, possesses a converging subsequence. \hfill $(H4)$

$$\nabla f(x) \neq \lambda \nabla g(x) \qquad \forall x \,:\, g(x) = b\,,\, f(x) = c\,,\, \forall \lambda > 0 \qquad (H5)$$

Then, for every $\bar{\varepsilon} > 0$ and every neighborhood $N \subset (\{f \leq c + \bar{\varepsilon}\} \cap \{g \leq b\})$ of $K_c \cap \{g \leq b\}$, there is a continuous $\eta : \{f \leq c + \bar{\varepsilon}\} \times [0,1] \to \{f \leq c + \bar{\varepsilon}\}$ and there is $\varepsilon \in (0, \bar{\varepsilon})$ such that:

$i)$ \qquad $\eta(x,s) = x$ if either $t = 0$, or $|f(x) - c| \geq \bar{\varepsilon}$, or $g(x) \geq b$

$ii)$ \qquad $\dfrac{\partial}{\partial s} f(\eta(x,s)) \leq 0$

$iii)$ \qquad $\eta(\{f \leq c + \varepsilon\} \cap \{g \leq b\}, 1) \subset \{f \leq c - \varepsilon\} \cap \{g \leq b\} \cup N$

$iv)$ \qquad $\eta(\{f \leq c + \varepsilon\} \cap \{g \leq b\} \setminus N, 1) \subset \{f \leq c - \varepsilon\} \cap \{g \leq b\}$

REMARK: If the critical set $K_c \cap \{g \leq b\} \cap \{g \leq b\}$ is empty, then the statement concludes as:

$$\eta(\{f \leq c + \varepsilon\} \cap \{g \leq b\}, 1) \subset \{f \leq c - \varepsilon\} \cap \{g \leq b\}\,.$$

In our setting the functionals f and g are respectively defined by

$$f(u) = \frac{1}{2} \int_0^T |\dot{u}|^2 - \int_0^T V(u,t) \qquad (4.1)$$

where V is the interaction potential as in (1.2);
f is defined and of class $\mathcal{C}^1$ on the domain

$$\Lambda = \{u = (u_1, \ldots, u_n) \in H^1_{T-per}(\mathbf{R}; \mathbf{R}^{kn}) \,:\, u_i(t) \neq u_j(t)\,, \forall t\,, \forall i \neq j\}\,,$$

and

$$g(u) = \frac{1}{2n} \sum_{i,j=1}^n \left(\frac{1}{T} \int_0^T u_i(t) - u_j(t) \right)^2. \qquad (4.2)$$

Here below we list some technical results from [26]. We refer to that paper for the proofs. Is is worthwhile to remark that Propositions 4.1, 4.2, 4.3, 4.4, 4.5 make Lemma 4.1 applicable to our functionals f and g for every values of c and b, with the bound $b > b(c)$, the function $b(\cdot)$ being defined in formula (4.3) below.

PROPOSITION 4.1. *Assume (V1) and (V2) hold. For every $c > 0$ there is $\delta(c) > 0$ such that*

$$f(u) \leq c \implies \min_{t \in \mathbf{R}} |u_i(t) - u_j(t)| \geq \delta(c) .$$

PROPOSITION 4.2. *If $g(u) > 0$ then $\nabla g(u) \neq 0$.*

PROPOSITION 4.3. *Assume (V1) and (V2) hold. For every $c, b \in \mathbf{R}$, every sequence $\{u_\nu\}_\nu \subset \Lambda$ such that $f(u_\nu) \to c$, $\nabla f(u_\nu) \to 0$ and $g(u_\nu) \leq b$ possesses a converging subsequence.*

PROPOSITION 4.4. *Assume (V1) and (V2) hold. For every $c \in \mathbf{R}$ and $b > 0$, every sequence $\{u_\nu\}_\nu \subset \Lambda$ such that $f(u_\nu) \to c$, and $g(u_\nu) \to b$ and $\nabla f(u_\nu) - \lambda_\nu \nabla g(u_\nu) \to 0$, with $\lambda_\nu \geq 0$, possesses a converging subsequence.*

PROPOSITION 4.5. *Assume (V1), (V2) and (V3) hold. For every $c > 0$ there exists $b = b(c) > 0$ such that the nonlinear eigenvalue problem*

$$\nabla f(u) = \lambda \nabla g(u), \quad \lambda > 0$$

has no solution u with

$$f(u) \leq c, \quad g(u) > b .$$

REMARK: More precisely we can take

$$b(c) = n R(n, \theta)^2 \left(\sqrt{\frac{cT}{6} \frac{1}{\cos \theta}} + \frac{\rho}{2} \right)^2 \tag{4.3}$$

where $R(n, \theta) = (c(\theta)^{n-1} - 1)(c(\theta) - 1)^{-1}$ if $\theta > 0$ and $R(n, 0) = n - 1$, and $c(\theta) = (1/2)(1 + (\cos \theta)^{-1})$.

PROPOSITION 4.6. *Let f and g as in (4.1), (4.2). We have*

$$i(\{u \in \Lambda : g(u) \geq n(n-1)^2 T f(u)/6\}) = 1 .$$

Let us define

$$\mathcal{A}_k = \{A \subseteq \Lambda \text{ closed} : i(A) \geq k\} \tag{4.4}$$

and

$$c_k^* = \inf_{A \in \mathcal{A}_k} \sup_A f. \tag{4.5}$$

Thanks to Corollary 3.1, for every $k \geq 1$, $\mathcal{A}_k \neq \emptyset$, indeed each class $\mathcal{A}_k$ contains at least one compact subset of Λ. Hence formula (4.5) really defines a real number for every $k \geq 1$. We remark that, by monotonicity of the sequence $\{\mathcal{A}_k\}_k$ with respect to k, there holds

$$c_k^* \leq c_{k+1}^* \qquad \forall k \geq 1.$$

For each $k \geq 1$, let us fix $\tilde{c}_k > c_k^*$ and let $b_k = b(\tilde{c}_k)$ as in (4.3), and define

$$c_k = \inf\{\gamma : (\{f \leq \gamma\} \cup (\{f \leq \tilde{c}_k\} \cap \{g \geq b_k\})) \in \mathcal{A}_k. \tag{4.6}$$

By definition, $\{f \leq c_k^* + \varepsilon\} \in \mathcal{A}_k$, for every $\varepsilon > 0$, hence

$$c_k \leq c_k^* < \tilde{c}_k, \qquad k = 1, 2, \ldots. \tag{4.7}$$

Now we claim that

$$c_k \geq c_{k-1}^*, \qquad k = 2, 3, \ldots. \tag{4.8}$$

Indeed, for every $\varepsilon > 0$ we have

$$i(\{f \leq c_k + \varepsilon\} \cup (\{f \leq \tilde{c}_k\} \cap \{g \geq b_k\})) \geq k$$

and hence, by subadditivity of i,

$$i(\{f \leq c_k + \varepsilon\}) \geq k - i(\{f \leq \tilde{c}_k\} \cap \{g \geq b_k\}) ;$$

since from (4.3) $b_k = b(\tilde{c}_k) \geq n(n-1)^2 T \tilde{c}_k / 6$, we have

$$\{f \leq \tilde{c}_k\} \cap \{g \geq b_k\} \subset \{u \in \Lambda : g(u) \geq n(n-1)^2 T f(u)/6\}$$

thus, by Proposition 4.6,

$$i(\{f \leq c_k + \varepsilon\}) \geq k - 1.$$

LEMMA 4.2. *Let $\{\tilde{c}_k\}_k$ be a sequence of real numbers such that $\tilde{c}_k > c_k^*$, $\forall k \geq 1$. Assume (V1)-(V3) hold: then, for each $k \geq 2$, f has at least one critical point u_k such that $f(u_k) = c_k$ and $g(u_k) \leq b(\tilde{c}_k)$*

PROOF: Suppose on the contrary that $K_{c_k} \cap \{g \leq b_k = b_k(\tilde{c}_k)\} = \emptyset$, for some $k \geq 2$. Then we obtain from Lemma 4.1 (which is applicable thanks of Propositions 4.1-5) that the set $\{f \leq c_k + \varepsilon\} \cup (\{f \leq \tilde{c}_k\} \cap \{g \geq b_k\})$ is deformable into the set $\{f \leq c_k - \varepsilon\} \cup (\{f \leq \tilde{c}_k\} \cap \{g \geq b_k\})$ in Λ, for small values of $\varepsilon > 0$. This fact contradicts the definition of the c_k, since, for $\varepsilon < \tilde{c}_k - c_k^*$ we have that

$$i(\{f \leq c_k + \varepsilon\} \cup (\{f \leq \tilde{c}_k\} \cap \{g \geq b_k\})) \geq k$$

while (for $k \geq 2$)

$$i(\{f \leq c_k - \varepsilon\} \cup (\{f \leq \tilde{c}_k\} \cap \{g \geq b_k\})) \leq k - 1$$

and each class $\mathcal{A}_k$ is invariant under homotopies of Λ into Λ (Proposition 2.1.ii)).

∎

LEMMA 4.3. *We have:*

$$\lim_{k \to +\infty} c_k^* = +\infty . \qquad (4.9)$$

PROOF : If not, we can assume that $c_k^* \to c^* = \sup_k c_k^*$, and we apply Lemma 4.2 with $\tilde{c}_k = \tilde{c}$, $\forall k$, where $\tilde{c}$ is any fixed real such that $\tilde{c} > c^*$. Then it follows from Lemma 4.2 and Proposition 4.3 that $K_{c^*} \cap \{g \leq \tilde{c}\}$ is a non-empty compact subset of Λ. We have, since $c^* \leq c_k^*$,

$$i(\{f \leq c^* + \varepsilon\}) = +\infty , \qquad \forall \varepsilon > 0 . \qquad (4.10)$$

Now let $k^* = i(K_{c^*} \cap \{g \leq b(\tilde{c})\})$ (which is a finite integer by Proposition 2.1.vi)), and let N be a neighborhood of $K_{c^*} \cap \{g \leq b(\tilde{c})\}$ such that $i(\overline{N}) = k^*$ as in Proposition 2.1.v). Using the subadditivity of the index i we obtain that, for every $\varepsilon < \tilde{c} - c^*$,

$$i(\{f \leq c^* + \varepsilon\}) \leq$$
$$i((\{f \leq c^* + \varepsilon\} \cap \{g \leq b(\tilde{c})\}) \setminus N) + i(\{f \leq c^* + \varepsilon\} \cap \{g \geq b(\tilde{c})\}) + i(N) ,$$

and hence, taking into account of Proposition 4.6 we obtain that

$$i((\{f \leq c^* + \varepsilon\} \cap \{g \leq b(\tilde{c})\}) \setminus N) \geq i(\{f \leq c^* + \varepsilon\}) - 1 - k^* .$$

Thus

$$i((\{f \leq c^* + \varepsilon\} \cap \{g \leq b(\tilde{c})\}) \setminus N) = +\infty$$

The application of Lemma 4.1 (taking into account of Proposition 2.1 parts ii) and iii)) yields

$$+\infty = i((\{f \leq c^* + \varepsilon\} \cap \{g \leq b(\tilde{c})\}) \setminus N)$$
$$\leq i(\eta(\{f \leq c^* + \varepsilon\} \cap \{g \leq b(\tilde{c})\} \setminus N), 1)$$
$$\leq i(\{f \leq c^* - \varepsilon\} \cap \{g \leq b(\tilde{c})\}) \leq k$$

for every k such that $c_k^* \geq c^* - \varepsilon$, a contradiction. ∎

PROOF OF THEOREM 1. It follows from Lemma 4.2 that each value c_k as defined in (4.6) is a critical level for f, if $k \geq 2$. Moreover, from Lemma 4.3 and the inequality (4.8), the sequence $\{c_k\}_k$ is unbounded. ∎

REFERENCES

[1] Ambrosetti A., Coti Zelati V. : *Critical points with lack of compactness and applications to singular dynamical systems.* Ann. di Matem. Pura e Appl. (Ser. IV) CIL (1987), 237-259.

[2] Ambrosetti A., Coti Zelati V. : *Noncollision orbits for a class of Keplerian-like potentials.* Ann. IHP Analyse non linéaire 5 (1988), 287-295.

[3] Ambrosetti A., Coti Zelati V. : *Periodic solutions for N-body type problems.* Ann. IHP Analyse non linéaire. to appear

[4] Bahri A., Rabinowitz P.H. : *A minimax method for a class of Hamiltonian system with singular potentials.* J. Funct. Anal, 82 (1989),412-428.

[5] Bahri A., Rabinowitz P.H. : *Periodic solutions of Hamiltonian systems of 3-body type.* Ann. IHP Anal. non Lin. 8 (1991), 561-649.

[6] Bessi U., Coti Zelati V. : *Symmetries and non-collision closed orbits for planar N-body type problems.* Preprint SISSA, (1990).

[7] Capozzi A., Solimini S., Terracini S. : *On a class of dynamical systems with singular potential.* Nonlin. Anal. TMA 16 (1991), 805-815.

[8] Coti Zelati V. : *Perturbations of Hamiltonian Systems with Keplerian Potentials.* Math. Zeits, 201 (1989), 227-242.

[9] Fadell E., Husseini S. : *A note on the category of the free loop space.* Proc. Amer. Math. Soc., 107 (1989), 527-536

[10] Fournier G., Willem M. : *Relative category and the calculus of variations.* Variational Problems (Ed. H.Berestycki, J.M.Coron, I.Ekeland). Birkhduser, Basel (1990).

[11] Degiovanni M., Giannoni F. : *Dynamical systems with Newtonian type potentials.* to appear in Ann. Scuola Norm. Sup Pisa Cl Sci (4) (1989).J

[12] Gordon W. : *Conservative dynamical systems involving strong forces.* Trans. AMS 204 (1975), 113-135.

[13] Greco C. : *Periodic solutions of a class of singular Hamiltonian systems.* Nonlin. Anal. TMA 12 (1988), 259-269.

[14] Majer P. : *Ljusternik-Schnirelman theory with local Palais-Smale condition and singular dynamical systems.* Ann. IHP Analyse non linéaire, to appear

[15] Majer P. : *Variational methods on manifolds with boundary.* Preprint SISSA 1991.

[16] Majer P., Terracini S. : *Periodic solutions to some n-body type Problems.* Quaderni del Dipartimento di Matematica del Politecnico di Milano, 1991.

[17] Meyer K.R. : *Periodic solutions of the N-body problem.* J. Diff. Eq., 39 (1981), 2-38.

[18] Moser J. : *Regularization of Kepler's problem and the averaging method on a manifold.* Comm. Pure Appl. Math., 23 (1970), 609-636.

[19] Serra E., Terracini S. : *Noncollision solutions to some three-body problems.* Preprint (1990). To appear in Arch. Rat. Mech. Anal.

[20] Spanier E. : *Algebraic Topology.* Mc Graw-Hill (1966).

THE ROLE OF THE DOMAIN SHAPE ON THE EXISTENCE AND MULTIPLICITY OF POSITIVE SOLUTIONS OF SOME ELLIPTIC NONLINEAR PROBLEMS

Giovanna Cerami
Facoltá di Ingegneria
Universitá di Palermo

1. INTRODUCTION

The aim of this note is to give the main ideas and to summarize the results of some investigations on the effect of the domain shape on the existence and multiplicity of positive solutions for some elliptic semilinear problems. The problems, that will be discussed, have the form

$$\begin{cases} -\varepsilon \Delta u = g(u) & \text{in } \Omega \\ u > 0 & \text{in } \Omega \\ u = 0 & \text{on } \partial\Omega \end{cases} \qquad (P_\varepsilon)$$

where $\Omega \subset \mathbb{R}^N$ is a smooth domain, $N \geq 3$, $\varepsilon \in \mathbb{R}^+ - \{0\}$ and $g : \mathbb{R} \longrightarrow \mathbb{R}$ is a $\mathscr{C}^{1,1}$ function. During the past few years the question if the geometry or topology of Ω can affect the solvability and the number of solutions of problem like (P_ε) has been object of several studies. The interest was firstly mainly related to variational problems with lack of compactness, i.e. problems in wich the usual techniques to prove the existence of solutions (minimization, critical point theory) cannot be applied directly. In particular the researchers attention was focused on the special case $g(u) = u^{2^*-1}$, $2^* = \frac{2N}{N-2}$, Ω bounded. Indeed the most relevant results on the existence problem for

$$\begin{cases} -\Delta u = u^{2^*-1} & \text{in } \Omega \\ u > 0 & \text{in } \Omega \\ u = 0 & \text{on } \partial\Omega \end{cases} \qquad (P_*)$$

have been, from the beginning, strongly connected to the shape of Ω. It is well known in fact that both, the celebrated Pohozaev nonexistence theorem [Po] and the first existence result by Kazdan-Warner [K.W] were obtained under geometric conditions on Ω : Ω starshaped the first, Ω an annulus the second. Nevertheless the origin of so different behaviour was not evident at all. More recently the remarkable studies by Coron [C] and Bahri-Coron [Ba.C] emphasized the effect of the domain topology and contributed to clarify the reasons of its role. In [Ba-C] the existence of a solution for (P_*) was proved under the assumption on Ω to be topologically nontrivial (in a suitable sense). The proof was based on the fact that it is possible to relate the change of topology of the sublevel sets, of a variational functional associated to (P_*), to the topology of Ω. On the other hand, almost at the same time, it was pointed out that the solvability of (P_*) cannot depend just on the domain topology. In fact in [P], [Di], [D.2], contractible domains, having special geometric properties, in which (P_*) has one or more solutions, were exhibited.

However, it would be restrictive to think that the domain shape plays a role only when the nonlinear term has a critical behaviour.

Of course, if Ω is bounded and g has a subcritical growth, under appropriate assumptions, for instance $g(t) = t^{p-1} - t$, $p \in (2, 2^*)$, the existence of a solution of (P_ε), is a well known fact. If Ω is a ball the uniqueness of the solution has been proved under various assumptions on g by Gidas-Ni-Nirenberg [G.N.N], Ni-Nussbaum [N.N], Kwong [K] in the superlinear case, by Brezis-Oswald [B.O] in the sublinear case. On the other hand in the paper by Brezis-Nirenberg [B.N], classical by now, it was observed, among other things, that if Ω is an annulus and $g(t) = t^{p-1} + \lambda t$, with $\lambda > 0$ near to 0 and p near to 2^*, (P_ε) has at least two solutions. Also Dancer, in [D.1], has obtained multiplicity of solutions to (P_ε) when $g(t) = t^{p-1}$, for special domains that are approximations of a finite union of disjoint balls. Then it is clear that, in the subcritical case, the intersting question is how the domain topology or geometry can affect the number of solutions of (P_ε). To this problem is devoted the attention in a series of papers [B.C.1], [B.C.2], [B.C.P], [C.P.1], [C.P.2], whose conclusions, that will be described in the following sections, indicate that the shape of Ω is relevant when the nonlinearity acts strongly, in some sense, on the problem. In this case concentration phenomena occur and it is possible to show that low energy sublevels of the variational functional associated to (P_ε) have topology as rich as Ω has. Problem (P_ε) has been studied either if Ω is bounded ([B.C], [B.C.P], [B.C.2]) either if Ω is unbounded ([C.P.1], [C.P.2]), using both Ljusternik-Schnirelman theory and Morse theory tools. The outline of this paper is the following: section 2 and 3 are devoted to the description of the results that it is possible to obtain when Ω is a not contractible bounded domain, using Ljusternik-Schnirelman theory (section 2) and Morse theory (section 3); section 4 contains the study of (P_ε) when Ω is unbounded as well as multiplicity results for contractible domains that are "perturbations" (in a suitable sense) of topologically rich domains.

2.
Throughout this section Ω will be a bounded smooth domain, also g is supposed to satisfy

$$g(0) = 0, \quad g'(0) < 0$$

so g can be written as

$$g(t) = g'(0)t + f(t) ,$$

and, putting, without any loss of generality, $g'(0) = -1$, (P_ε) takes the form

$$\begin{cases} -\varepsilon \Delta u + u = f(u) & \text{in } \Omega \\ u > 0 & \text{in } \Omega \\ u = 0 & \text{on } \partial\Omega \end{cases} \tag{2.1}$$

where f is, by definition, a $\mathcal{C}^{1,1}$ function such that

$H_1)$ $\quad f(0) = f'(0) = 0$.

Moreover f is assumed to verify the following hypotheses

$H_2)$ $\quad$ there exists a $\in \mathbb{R}^+ - \{0\}$, such that, for every $t \in \mathbb{R}$

$$|f(t)| \le a + a\, t^p$$

$$|f'(t)| \le a + a\, t^{p-1}$$

where $p \in \left(1, \dfrac{N+2}{N-2}\right)$

$H_3)$ $\quad$ there exists $\vartheta \in (0, 1/2)$ such that

$$F(t) \le \vartheta t\, f(t) \qquad \forall\, t \in \mathbb{R}$$

where

$$F(t) = \int_0^t f(s)\, ds \qquad t \in \mathbb{R}$$

$H_4)$ $\quad \dfrac{d}{dt}\left[\dfrac{f(t)}{t}\right] > 0 \qquad \forall\, t > 0$.

Also in what follows the condition

$H_5)$ $\quad f(t) = 0 \qquad t < 0$

is supposed to be satisfied.

Remark 2.1 Note that $H_5)$ is not restrictive, in fact, if it were not true, problem (2.1) could be replaced by the equivalent

$$\begin{cases} -\varepsilon\, \Delta u + u = f^*(u) & \text{in } \Omega \\ u = 0 & \text{on } \partial\Omega \end{cases} \qquad (2.2)$$

where

$$f^*(t) = \begin{cases} f(t) & t \ge 0 \\ 0 & t < 0 \end{cases}.$$

The equivalence of (2.1) and (2.2) is a consequence of the maximum principle and of (H_2) that implies $f^*(t) = f(t) \geq kt^{1/\vartheta - 1}$, $t > 0$ with $k > 0$ constant.

The solutions of (2.1) are the nontrivial critical points of the energy functional

$$E_\varepsilon : H^1_0(\Omega) \longrightarrow \mathbb{R}$$

defined by

$$E_\varepsilon(u) = \frac{1}{2} \int_\Omega \left(\varepsilon |\Delta u|^2 + u^2 \right) dx - \int_\Omega F(u) \, dx \; . \tag{2.3}$$

It is well known that under the previous assumptions (2.1) has a (positive) solution that can be found, for instance, applying the Mountain-Pass theorem [A.R] to E_ε.

Using the Ljusternik-Schnirelman theory, it is possible to establish the following results

<u>Theorem 2.2</u> <u>Suppose f satisfies the assumptions H_1), ..., H_5) and Ω is not contractible, then there exists $\varepsilon^* > 0$ such that, for any $\varepsilon \in (0, \varepsilon^*]$, (2.1) has at least (cat Ω)+1</u>[1] <u>distinct solutions.</u>

<u>Theorem 2.3</u> <u>Suppose $f(t) = t^{p-1}$, Ω not contractible, then for any $\varepsilon \in \mathbb{R}^+ - \{0\}$, there exists $\bar{p} = \bar{p}(\varepsilon) \in (2, 2^*)$ such that $\forall p \in [\bar{p}, 2^*)$ problem (2.1) has at least (cat Ω)+1 distinct solutions.</u>

Theorem 2.2, in the above form, for general nonlinearities, has been proved in [B.C.2]. Theorem 2.3, as well Theorem 2.2 for the case $f(t) = t^{p-1}$, have been stated first in [B.C.1] with the less general conclusion of the existence of (cat Ω) distinct solutions. Then the results have been improved, showing the existence of at least one more solution, in [B.C.P]. We notice also that in [B.C.1] and [B.C.P] the equation considered was written in the form $-\Delta u + \lambda u = u^{p-1}$ that,

---

[1] cat Ω denotes the Ljusternik-Schnirelman category of $\bar{\Omega}$ in itself.

because of the homogeneity of the nonlinear part is equivalent to $-\varepsilon\Delta u + u = u^{p-1}$, $\varepsilon = 1/\lambda$[it is, in fact, easy to verify that to any solution u of the last one there corresponds, by the translation $v = (1/\varepsilon)^{1/p-1} u$, a solution of the first one with $\lambda = 1/\varepsilon$.

The conclusions of the above theorems can be summarized in this way: if Ω has "rich" topology, (1.1) has many solutions when the nonlinearity acts strongly on the problem. In these cases, in fact, concentration phenomena appear. Indeed, it is well known that concentration is a typical effect connected to the critical growth of the nonlinearity, but it occurs also, for any fixed $p \in (2,2^*)$, when ε is small, because, after a suitable rescaling, problem (2.1) looks like an associated problem in the whole space. In fact a minimal energy solution of (2.1), when ε is small, is very similar to a ground state solution of (2.1), with $\varepsilon = 1$, in $\mathbb{R}^N$, highly concentrated around some point of Ω.

The idea of the proof of Theorems 2.2 and 2.3 is very simple: to show that it is possible to relate the topology of Ω to the topology of some sublevel sets of a variational functional connected to the problem, then apply classical results of the Ljusternik-Schnirelman theory, that give a lower bound to the number of critical points of a functional on a manifold in terms of some topological invariant of the manifold.

We sketch just the main steps of the proof of Theorem 2.2 (for more details see [B.C.2]), Theorem 2.3 can be proved with the same arguments with obvious modifications and simplifications (see [B.C.1] and [B.C.P]).

The solutions of (2.1), as before said, are nontrivial critical points of the functional E_ε defined in $H_0^1(\Omega)$ by (2.3), but, as well known, this functional is not easy to handle, because it is bounded neither from below nor from above. So, first of all, it is useful to replace E_ε with another functional, having the same critical points, but easier to work with.

Let consider the smooth manifold

$$\mathcal{S}_\Omega = \left\{ v \in H_0^1(\Omega) : \|v\|_{H_0^1} = 1 \right\} \setminus \left\{ v \in H_0^1(\Omega) : v \leq 0 \text{ a.e.} \right\}$$

and consider for any $v \in \mathcal{S}_\Omega$

$$I_\varepsilon(v) = \sup_{\xi \in \mathbb{R}^+} E_\varepsilon(\xi v) . \tag{2.4}$$

The assumptions $H_1)...H_3)$ imply that the supremum in (2.4) is achieved, $H_4)$ guarantees that for any $v \in \mathcal{S}_\Omega$, there is just one real positive value $\xi = \xi(v)$ such that

$$I_\varepsilon(v) = E_\varepsilon\bigl(\xi(v)v\bigr) = \max_{\xi \in \mathbb{R}^+} E_\varepsilon(\xi v) .$$

So I_ε is a well defined functional on $\mathcal{S}_\Omega$, that, by the regularity properties of f, turns out to be $\mathcal{C}^{1,1}$. Also using $H_1)...H_4)$ it is not difficult to verify that I_ε enjoys of the following properties

i) $I_\varepsilon(v) \geq k_\varepsilon > 0 \quad \forall v \in \mathcal{S}_\Omega, \ k_\varepsilon \equiv \text{constant}$

ii) I_ε satisfies P.S. on $\mathcal{S}_\Omega$ i.e. if $\{v_n\} \subset \mathcal{S}_\Omega$ is such that $I_\varepsilon(v_n)$ is bounded and $\|I'_\varepsilon(v_n)\| \longrightarrow 0$ then v_n has a subsequence convergent in $\mathcal{S}_\Omega$.

iii) u is nontrivial critical point of E_ε on $H^1_0(\Omega)$ if and only if $v = \frac{u}{\|u\|}$ is a critical point of I_ε on $\mathcal{S}_\Omega$.

iv) the infimum

$$\inf \left\{ I_\varepsilon(v) , \ v \in \mathcal{S}_\Omega \right\} = m(\varepsilon,\Omega)$$

is achieved, moreover to this infimum there corresponds the positive solution of (2.1) found applying the mountain pass theorem to E_ε.

Now, for $c \in \mathbb{R}$, denote by

$$I^c_\varepsilon = \left\{ v \in \mathcal{S}_\Omega : I_\varepsilon(v) \leq c \right\}$$

and observe that, if it is possible to show that for some $c \in \mathbb{R}, \ c > 0$

$$\text{cat}_{I^c_\varepsilon} I^c_\varepsilon \geq \text{cat}_{\overline{\Omega}} \overline{\Omega} , \tag{2.5}$$

then the existence of at least cat Ω solutions to (2.1) turns out as a consequence of the following classical theorem, whose hypotheses are fulfilled by the couple $(I_\varepsilon, I_\varepsilon^c)$:

Theorem 2.4 Let h be a $\mathcal{C}^{1,1}$ functional, defined on a smooth manifold M, bounded from below and satisfying P.S. on it, then h has on M at least cat M critical points.

To carry out the program of establishing (2.5), for some energy value c of I_ε, the abstract tool is the following :

Proposition 2.5 Given H, Ω^- and Ω^+ closed sets with $\Omega^- \subset \Omega^+$ and two continuous maps

$$\beta : H \longrightarrow \Omega^+$$
$$\psi : \Omega^- \longrightarrow H$$

such that $\beta \circ \psi$ is homotopically equivalent to the embedding $j : \Omega^- \longrightarrow \Omega^+$, then

$$\operatorname{cat}_H H \geq \operatorname{cat}_{\Omega^+} \Omega^-.$$

The proof of this proposition (see [B.C.1] or [B.C.P]) can be easily done just applying the definitions of category and of homotopyc equivalence between maps and showing that, if n is the smallest positive integer such that $H \subset \bigcup_{i=1}^{n} A_i$, A_i closed contractible sets in H, then the sets $K_i \equiv \psi^{-1}(A_i)$, i=1,2,...,n, are contractible sets in Ω^+ and cover Ω^-.

In order to apply Proposition 2.5, let define

$$\Omega^+ = \{ x \in \mathbb{R}^N : d(x,\Omega) \leq r \}$$

$$\Omega^- = \{ x \in \Omega : d(x,\partial\Omega) \geq 2r \}$$

and chose r so small that Ω^+ and Ω^- are topologically equivalent to Ω, so

$$\operatorname{cat}_{\Omega^+} \Omega^- = \operatorname{cat}_{\overline{\Omega}} \overline{\Omega}.$$

Then consider

$$c \equiv c_\varepsilon = m(\varepsilon, r)$$

where for $\rho > 0$

$$m(\varepsilon, \rho) = \inf \left\{ I_\varepsilon(v), v \in \mathscr{S}_{B_\rho(0)} \right\} \quad \text{and}$$

$$\mathscr{S}_{B_\rho(0)} = \left\{ v \in H_0^1(B_\rho(0)) : \|v\|=1 \right\} \setminus \left\{ v \in H_0^1(B_\rho(0)) : v \leq 0 \text{ a.e.} \right\}$$

and define the maps

$$\beta : H_0^1(\Omega) \longrightarrow \mathbb{R}^N$$

by

$$\beta(u) = \frac{\int_\Omega x \, |\nabla u|^2 \, dx}{\int_\Omega |\nabla u|^2 \, dx}.$$

$$\Psi : \bar\Omega \longrightarrow \mathscr{S}_\Omega$$

by

$$\Psi(y) \equiv \Psi_\varepsilon(y) = \begin{cases} u_\varepsilon(y-x) & x \in B_{2r}(y) \\ 0 & x \notin B_{2r}(y) \end{cases}$$

where u_ε is the positive function, radially symmetric about the origin (see [G.N.N]) that realizes $m(\varepsilon, 2r)$.

It is clear that, for any choice of ε, β and Ψ are continuous maps, that

$$(\beta \circ \Psi)(y) = y \qquad \forall \, y \in \bar\Omega$$

and that $\forall \, y \in \bar\Omega$

$$I_\varepsilon(\Psi(y)) = m(\varepsilon, 2r) < m(\varepsilon, r)$$

so

$$\Psi(\overline{\Omega}^-) \subset I_\varepsilon^{m(\varepsilon,r)} .$$

Thus the last relation to verify, in order to get (2.5), via Proposition 2.5 (with $H = I_\varepsilon^{m(\varepsilon,r)}$), is

$$\beta \left(I_\varepsilon^{m(\varepsilon,r)} \right) \subset \Omega^+ . \qquad (2.6)$$

This is a very delicate point, in fact (2.6) is not true in general, and the choice of ε plays a decisive role. The argument of the proof is tricky and technically complex, here we try to explain the underlying idea. The scale change

$$\sigma(u(x)) = u(\sqrt{\varepsilon}\, x)$$

shows the function that realizes

$$m(\varepsilon, \mathbb{R}^N) \equiv \inf \left\{ I_\varepsilon(v) , v \in \mathcal{S}_{\mathbb{R}^N} \right\}$$

$$\mathcal{S}_{\mathbb{R}^N} = \left\{ u \in H^1(\mathbb{R}^N) , \|u\|_{H^1(\mathbb{R}^N)} = 1 \right\} \setminus \left\{ u \in H^1(\mathbb{R}^N) : u \leq 0 \text{ a.e.} \right\}$$

is nothing but the function that realizes

$$m(1,\mathbb{R}^N) \equiv \inf \left\{ I_1(v) , v \in \mathcal{S}_{\mathbb{R}^N} \right\}$$

re-scaled and the more concentrated about some point in $\mathbb{R}^N$ the more ε is small. So it is not hard to realize that if ε is small enough, $m(\varepsilon,\mathbb{R}^N)$, $m(\varepsilon,\Omega)$, $m(\varepsilon,r)$ are values very close each other, and to imagine that a function that realizes $m(\varepsilon,r)$ [$m(\varepsilon,\Omega)$] looks like the function that realizes $m(\varepsilon,\mathbb{R}^N)$ concentrated about the center of B_r [about some point of Ω] and cut off outside of B_r [Ω]. Also one can reasonably suppose, if ε is small enough, that the shape of the functions $v \in \mathcal{S}_\Omega$, having energy $I_\varepsilon(v) \leq m(\varepsilon,r)$ is similar to that, above described, of the minimization functions and to understand that (2.6) is true.

The last step to achieve the proof of Theorem 2.2 is to show that, considering energy levels γ higher than $m(\varepsilon,r)$, at some point the topological type of I_ε^γ changes, and this implies, by the Ljusternik-Schnirelman theory, the existence of one more stationnary point of I_ε. The geometric idea is once again very simple. Consider the set $\Gamma = \overline{\Psi(\Omega)}$ this set is not contractible. Taken $v^* \in \mathcal{S}_\Omega$ such that $v^* \notin \Gamma$ define the "cone"

$$\Theta = \{ \vartheta v^* + (1-\vartheta)v , v \in \Gamma , \vartheta \in [0,1] \} .$$

Θ is compact and contractible to the point v^*, $0 \notin \Theta$, hence the set

$$\Lambda = \left\{ \frac{w}{\|w\|} : w \in \Theta \right\}$$

is well defined, and

$$\Gamma \subset \Lambda \subset \mathcal{S}_\Omega .$$

Putting

$$\gamma_\varepsilon = \max \{ I_\varepsilon(w) , w \in \Lambda \}$$

Λ, and then Γ turns out to be contractible in $I_\varepsilon^{\gamma_\varepsilon}$, this means that there is a change of topology between the levels $m(\varepsilon,r)$ and γ_ε, as wanted.

We close this section mentioning that it has been possible to apply the ideas before exposed also to the study of the problem

$$\begin{cases} -\Delta u - \lambda u = u^{2^*-1} & \text{in } \Omega \\ u > 0 & \text{in } \Omega \\ u = 0 & \text{on } \partial\Omega \end{cases} \qquad (2.7)$$

In fact first Lazzo in [L] has proven that if $\Omega \subset \mathbb{R}^N$, $N \geq 4$, is bounded, for $\lambda > 0$ small enough, (2.7) has at least (cat Ω) solutions, improving previous results by Rey [R] obtained using the Green function properties. More recently Passaseo [P.2] has shown that, actually, the solutions of (2.7) are at least (cat Ω)+1.

We mention also that some multiplicity results for problem like (2.1) have been obtained in [Ca].

3.

This section is devoted to the description of the results that it is possible to obtain using the Morse theory. In order to do this some notations and facts about the generalized Morse theory (see f.i. [B] or [Ch] are needed.

If (X,Y) is a couple of topological spaces set

$$P_t(X,Y) = \sum_k \dim \left[H_k(X,Y) \right] t^k$$

where $H_k(X,Y)$ is the k-th homology group with coefficients in some field, and

$$P_t(X) = P_t(X,\phi).$$

$P_t(X)$ is called the Poincaré polynomial of X.

Let M be a smooth Hilbert-Riemannian manifold and let h a $\mathcal{C}^{1,1}$ function defined on M. If u is an isolated critical point of h on M, such that $h(u) = c$, its polynomial Morse index is

$$i_t(u) = P_t(h^c \cap U, (h^c \setminus \{u\}) \cap U) =$$
$$= \sum_k \left[\dim H_k(h^c \cap U, (h^c \setminus \{u\}) \cap U) \right] t^k$$

where

$$h^c = \{ u \in M : h(u) \le c \}$$

and U is a neighbourhood of u such that, the only critical point of h belonging to $h^c \cap U$ is u.

The integer number $i_1(u)$ is called the multiplicity of u. If h is twice differentiable at u, and the Hessian $d^2h(u)$ has bounded inverse, u is called nondegenerate. Since $A = d^2h(u)$ is a self-adjoint operator which possesses a resolution of identity, we call the dimension of the negative space corresponding to the spectral resolution, the (numerical) Morse index of u. If u is a nondegenerate critical point of h, whose Morse index is $\mu(u)$, then

$$i_t(u) = t^{\mu(u)}.$$

Clearly if u is nondegenerate, its multiplicity is 1. If the multiplicity is n, a generic $\mathscr{C}^2$-perturbation splits u into n nondegenerate solutions.

The main result that it has been possible to state about problem 2.1 using Morse theory is :

<u>Theorem 3.1 [B.C.2]</u> <u>Suppose that f satisfies the assumptions $H_1)...H_5)$. Moreover suppose that</u>

i) <u>$\varepsilon \in (0, \varepsilon^*]$ with ε^* suitable positive constant</u>
ii) <u>the set $\mathcal{K}_\varepsilon$ of the nontrivial solutions of problem 2.1 is discrete; then</u>

$$\sum_{u \in \mathcal{K}_\varepsilon} i_t(u) = t\, P_t(\Omega) + t^2\left[P_t(\Omega) - 1\right] + t(1+t)\, Q(t) \qquad (3.1)$$

<u>where Q(t) is a polynomial with non-negative integer coefficients.</u>

Let notice that Theorem 3.1 implies that problem (2.1) has at least $2P_1(\Omega)-1$ solutions if they are counted with their multiplicity. Of course if Ω is topologically trivial, then $P_1(\Omega)=1$, and the above theorem does not give any extra information. But, when Ω is topologically rich, it allows to obtain good informations. An example of this is given by the following

<u>Corollary 3.2 [B.C.2]</u> <u>Let A and C_i, (i=1,2...,k) be contractible open non empty sets in $\mathbb{R}^N$, smooth and bounded; suppose that</u>

$$\overline{C}_i \cap \overline{C}_j = \phi \qquad i,j = 1,2,...,k \quad ; \quad i \neq j$$

$$\overline{C}_i \subset A \qquad i = 1,2,...,k$$

and set

$$\Omega = \mathcal{A} \setminus (\bigcup_{i=1}^{k} \overline{C}_i)$$

Then there exists $\varepsilon^* > 0$, such that, for any $\varepsilon \in (0, \varepsilon^*]$, problem 2.1 has at least 2k+1 solutions, if they are counted with their multiplicity.

Moreover, if the solutions are non degenerate, k of them have Morse index N, k of them have Morse index N+1, and one (the Mountain Pass solution) has index 1.

The above statement is a straight consequence of theorem 3.1 and of the fact that an easy calculation of algebraic topology gives $P_t(\Omega) = 1 + kt^{N-1}$. It emphasizes also the advantage that Morse theory can give with respect to the Ljusternik-Schnirelman theory. Results analogous can be proved if the nonlinearity is of polynomial type and has a growth near to the critical one :

__Theorem 3.3__ Suppose $f(t) = t^{p-1}$ and let the assumption (ii) of Theorem 3.1 be satisfied. Then for any $\varepsilon > 0$ there exists $\overline{p} = \overline{p}(\varepsilon) \in (2, 2^*)$, such that if $p \in [\overline{p}, 2^*)$ the relation (3.1) holds.

The proof of Theorem 3.1, that is contained in [B.C.2], is carried out studying directly the topology of the sublevel sets of E_ε. However this needs the use of strong tools of algebraic topology. If the assumption (ii) in the Theorem 3.1 is replaced by the

ii)' the solutions of (2.1) are non degenerate

then the proof can be done in a simpler way, using the Morse theory for the functional I_ε. Here a sketch of this proof will be given.

In the next, if $\overline{u}$ is a non degenerate critical point of E_ε, $\mu(\overline{u})$ will denote its numerical Morse index, while $\nu(\overline{v})$ will be the numerical Morse index of $\overline{v} = \dfrac{\overline{u}}{\|\overline{u}\|}$ as critical point of I_ε.

If ii)' is assumed, the relation 3.1 becomes

$$\sum_{u \in \mathcal{K}_\varepsilon} t^{\mu(u)} = t\, P_t(\Omega) + t^2\bigl[P_t(\Omega) - 1\bigr] + t(1+t)\, Q(t). \tag{3.2}$$

Called $\mathcal{C}_1$ and $\mathcal{C}_2$ the sets of critical points of I_ε having energy less or equal than $m(\varepsilon,r)$, bigger than $m(\varepsilon,r)$ respectively, the proof of (3.2) is done showing that:

$$\sum_{v \in \mathcal{C}_1} t^{\nu(v)} = P_t(\Omega) + Z(t) + (1+t)\, Q_1(t). \tag{3.3}$$

$$\sum_{v \in \mathcal{C}_2} t^{\nu(v)} = t\left[P_t(\Omega) + Z(t) - 1 \right] + (1+t)\, Q_2(t), \tag{3.4}$$

where $Z(t)$, $Q_1(t)$ and $Q_2(t)$ are polynomials with nonnegative integer coefficients, and then proving that if $\bar{u}$ is a critical point of I_ε and $\bar{v}$ is the corresponding critical point of I_ε on $S_{\Omega'}$

$$\mu(\bar{u}) = \nu(\bar{v}) + 1. \tag{3.5}$$

The relation (3.5) can be verified just applying the definition of numerical Morse index. An intuitive understanding of it can be found observing that $\mathcal{S}_\Omega$ is a manifold of codimension 1.

The equality (3.3) is a direct consequence of the argument developed in the first part of the proof of Theorem 2.2. In fact denoted by Ψ_k and β_k the homomorphisms induced by Ψ and β, respectively, between the k-th homology groups, it turns out $\beta_k \circ \Psi_k = \mathrm{Id}_k$, $H_k(\Omega^-)$ homotopic to a subspace of $H_k(I_\varepsilon^{m(\varepsilon,r')})$ and

$$\dim(H_k(\Omega)) = \dim(H_k(\Omega^-)) \leq \dim H_k(I_\varepsilon^{m(\varepsilon,r)}).$$

Hence

$$P_t(I_\varepsilon^{m(\varepsilon,r)}) = P_t(\Omega) + Z(t), \tag{3.6}$$

with $Z(t)$ polynomial with nonnegative integer coefficients. On the other hand, since I_ε satisfies the Palais-Smale condition, from the Morse theory, it follows

$$\sum_{v \in \mathcal{C}_1} t^{\nu(v)} = P_t(I_\varepsilon^{m(\varepsilon,r)}) + (1+t) Q_1(t) \qquad (3.7)$$

with Q_1 polynomial with nonnegative integer coefficients. Hence (3.3) is obtained joining (3.6) and (3.7).

The proof of (3.4) needs to computing $P_t(\mathscr{S}_\Omega, I_\varepsilon^{m(\varepsilon,r)})$, because by the Morse theory

$$\sum_{v \in \mathcal{C}_1} t^{\nu(v)} = P_t(\mathscr{S}_\Omega, I_\varepsilon^{m(\varepsilon,r)}) + (1+t) Q_2(t) \qquad (3.8)$$

where Q_2 is a polynomial with nonnegative integer coefficients. Looking at the exact sequence

$$\longrightarrow H_k(\mathscr{S}_\Omega) \xrightarrow{j_k} H_k(\mathscr{S}_\Omega, I_\varepsilon^{m(\varepsilon,r)}) \xrightarrow{\partial_k} H_{k-1}(I_\varepsilon^{m(\varepsilon,r)}) \xrightarrow{i_{k-1}} H_{k-1}(\mathscr{S}_\Omega)$$

it is not difficult to verify that

$$\dim\left[H_k(\mathscr{S}_\Omega, I_\varepsilon^{m(\varepsilon,r)})\right] = \begin{cases} \dim\left[H_{k-1}(I_\varepsilon^{m(\varepsilon,r)})\right] & k \geq 2 \\ \dim j_1(H_1(\mathscr{S}_\Omega)) = 0 & k = 1 \end{cases}$$

Thus, since $H_0(\mathscr{S}_\Omega, I_\varepsilon^{m(\varepsilon,r)}) = 0$,

$$P_t(\mathscr{S}_\Omega, I_\varepsilon^{m(\varepsilon,r)}) = t\left[P_t(I_\varepsilon^{m(\varepsilon,r)}) - 1\right] = t\left[P_t(\Omega) + \mathcal{Z}(t) - 1\right],$$

so (3.4) is verified and the claim is proved.

4.

This section contains a brief description of the contributions given in [B.C.P.], [C.P.1] and [C.P.2] to two questions :

a) to understand better how the existence of multiple solutions of (2.1) (that has been shown strongly related to the topology of Ω) can appear also in some topologically trivial domains (as shown for instance in [D.1]);

b) to prove existence and multiplicity results also for problem (2.1) when Ω is an unbounded domain.

In what follows we shall be concerned with (2.1) when f has a polynomial growth, namely

$$\begin{cases} -\varepsilon \Delta u + u = u^{p-1} & \text{in } \Omega \\ u > 0 & \text{in } \Omega \\ u = 0 & \text{on } \partial\Omega \end{cases} \quad (4.1)$$

With regard to the first question, even if the results exposed in the previous sections seem to emphasize the role of the topology of Ω, the proof method suggests that the basic point is the topology of suitable nearby domains. Indeed the lower bound to the number of solutions to (4.1), if Ω is bounded, is given using $\text{cat}_{\Omega^+}\Omega^-$, with $\Omega^- \subset \Omega^+$ related to Ω in some sense. Now if Ω is topologically rich it is advantageous to choose Ω^- and Ω^+ equivalent to Ω, but, if Ω is a topologically trivial domain "near", in a suitable sense, to a rich domain Ω^*, it is clear that it is more convenient to try to construct Ω^+ and Ω^- topologically equivalent to Ω^*.

This is precisely the way in which some "perturbations" results have been established in [B.C.P.]. In order to state these we need to remind the notion of capacity of a set and to define an other useful quantity.

Definition 4.1 Let $\mathcal{D} \subset \mathbb{R}^N$ be a bounded smooth domain and let C be a nonempty subset of $\mathcal{D}$. If the set $\{u \in H^1(\mathcal{D}) : u \geq 1 \text{ on } C \text{ in } H^1(\mathcal{D}) \text{ sense}\}$ is nonempty, the capacity of C with respect to $\mathcal{D}$, $\text{cap}_\mathcal{D} C$, is the following number :

$$\text{cap}_\mathcal{D} C \equiv \inf\left\{\int_\mathcal{D} |\nabla u|^2 dx : u \in H^1_0(\mathcal{D}), u \geq 1 \text{ on } C \text{ in } H^1 \text{ sense}\right\}. \quad (4.2)$$

Moreover if $C = \phi$, $\text{cap}_\mathcal{D} \phi \equiv 0$.

Definition 4.2 Let L and $\mathcal{D}$ be two bounded smooth domains, we define

$$\mu(L,\mathcal{D}) = \inf\left\{\int_{\mathcal{D}\cup L}|\nabla u|^2 dx : u \in H_0^1(\mathcal{D} \cup L), \int_L |u|^p = 1\right\} \quad (4.3)$$

if the set

$$\left\{u \in H_0^1(\mathcal{D} \cup L) : \int_L |u|^p dx = 1\right\}$$

is nonempty, $\mu(L,\mathcal{D}) = +\infty$ otherwise.

Then the following theorems hold:

__Theorem 4.3__ Let $\Omega^* \subset \mathbb{R}^n$ be a bounded smooth domain not contractible. Let $R \in \mathbb{R}^+$ be such that $\overline{\Omega}^* \subset B_R(0) = \{x \in \mathbb{R}^N : |x| < R\}$. Let $K \subset \overline{\Omega}^*$ be a closed set. For any fixed $p \in (2, 2^*)$, there exists $\overline{\varepsilon} = \overline{\varepsilon}(p, \Omega^*) > 0$ such that for any $\varepsilon \in (0, \overline{\varepsilon}]$ it can be found $\delta = \delta(\varepsilon)$ for which, if $(\text{cap}_{B_R(0)} K) < \delta$, then problem (4.1) has in $\Omega = \Omega^* \setminus K$ at least $(\text{cat }\Omega^*)+1$ distinct solutions.

__Theorem 4.4__ Let Ω^*, R, K be as in Theorem 4.3. For any fixed $\varepsilon > 0$, there exists $\overline{p} = \overline{p}(\varepsilon, \Omega^*) \in (2, 2^*)$ such that for each $p \in [\overline{p}, 2^*)$ it can be found $\delta = \delta(p, \Omega^*) > 0$ for which, if $(\text{cap}_{B_R(0)} K) < \delta$, then the same conclusion of Theorem 4.3 holds.

__Theorem 4.5__ Let $\Omega^* \subset \mathbb{R}^n$ and $L \subset \mathbb{R}^n$ be bounded smooth domains; let Ω^* be not contractible. For any fixed $p \in (2, 2^*)$ there exists $\overline{\varepsilon} = \overline{\varepsilon}(p, \Omega^*) > 0$, such that for each $\varepsilon \in (0, \overline{\varepsilon}]$ it can be found $h = h(\varepsilon) > 0$ for which, if $\mu(L, \Omega^*) > h$, then problem (4.1) in $\Omega = \Omega^* \cup L$ has at least $(\text{cat }\Omega^*)+1$ distinct solutions.

__Theorem 4.6__ Let Ω^* and L be as in theorem 4.5. For any fixed $\varepsilon > 0$, there exists $\overline{p} = \overline{p}(\varepsilon, \Omega^*)$ such that for each $p \in [\overline{p}, 2^*)$ it can be found $h = h(p) > 0$ for which, if $\mu(L, \Omega^*) > h$, then the same conclusion of Theorem 4.5 holds.

The above results say that if we perturb a domain Ω^* cutting off or adding a set, small in a suitable sense, what is relevant, when the nonlinearity acts strongly on the problem, is still the topology of Ω^*. So, for instance, from Theorems 4.3 - 4.4 it can be straightforth deduced that in a contractible domain "near" to an annulus, (4.1) has at least three solutions if ε is small or p is near to 2^*, and the same results follows from Theorems 4.5 - 4.6 when Ω is a "dumb-bell" with a suitable thin, but not necessarily short, handle joining two balls.

The properties of the capacity are well known (see for ex. [S]), so it is clear what kind of sets it is possible to cut off from a nontrivial domain, saving the solutions multiplicity given by the topological richness. The quantity $\mu(L,\mathcal{D})$ has been defined in [B.C.P.] and, among other things, enjoys of the property expressed by

$$\mu(L,\mathcal{D}) \geq S/(\text{mis } L)^{2(2^*-p)/2^* p}$$

where S is the best Sobolev constant. So the conditions on L requested in Theorems 4.5 and 4.6 are verified if (mis L) < δ, δ suitably small. However this is not necessary and it is not hard to understand that a sufficient condition to have $\mu(L,\mathcal{D})$ big is that at least one of the dimension of L is small [either thickness, either lenght ...].

Let consider now (4.1), when Ω is an unbounded domain. In this case the lack of compactness of the embedding $j : H^1_0(\Omega) \longrightarrow L^p(\Omega)$ gives rise to a failing of the Palais-Smale condition and this is an hard obstacle when one wants to use variational techniques. First existence results were obtained using symmetry assumptions on Ω to overcome these difficulties (see [Be.L] for $\Omega = \mathbb{R}^n$, [E.L.] for the case in which Ω is the complement of a ball and [C.M.] for unbounded domains with other kinds of symmetry). Subsequently a careful analysis of the compactness question (see for ex. [B.C.3]) allowed to understand better the nature of the obstructions to the compactness. So an estimate of the energy levels where the Palais-Smale condition fails was possible and existence results for (4.1) were obtained for exterior domains which do not enjoy of any symmetry (see [B.C.3] for a first existence theorem under smallness assumptions on the complement of Ω, and [Ba.L.] for subsequent general results). In [C.P.1] and [C.P.2] problem (4.1) when Ω unbounded has been studied, with the purpose of connecting the existence and multiplicity of its solutions with the topology of the domain. Clearly, the peculiarity of the case Ω unbounded needs more complex topological and analytical machinery. The category does not seems the right topological invariant, and a lower bound to the number of solutions can be given in terms of the relative category. Before stating the theorem that summarizes these researches, we recall the definition of relative category.

<u>Definition</u> <u>Let X be a topological space, and X_1, X_2 closed subsets of X such that $X_2 \subseteq X_1$. We say that the relative category in X of X_1 with respect to X_2 is n and we write</u>

$$\text{cat}_X [X_1, X_2] = n$$

<u>if and only if n is the least nonnegative integer such that</u>

$$X_1 = \bigcup_{i=0}^{n} \mathcal{A}_i \qquad X_2 \subset \mathcal{A}_0$$

<u>where for each i, $\mathcal{A}_i$ is closed and there exists $h_i \in \mathcal{C}\bigl([0,1] \times \mathcal{A}_i, X\bigr)$ such that</u>

I) $h_i(0,x) = x$ $\forall\ x \in \mathcal{A}_i$, $i = 0,1,2,\ldots,n$
II) $\forall\ i \geq 1$ $\exists\ p_i \in X$: $h_i(1,x) = p_i$ $\forall\ x \in \mathcal{A}_i$.
III) $h_0(1,x) \in X_2$ $\forall\ x \in \mathcal{A}_0$
       $h_0(t,x) \in X_2$ $\forall\ x \in \mathcal{A}_0 \cap X_2$ $\forall\ t \in [0,1]$.   □

If we put

$$\bar{\rho} = \inf \left\{ \rho \in \mathbb{R} \ :\ \mathbb{R}^N \setminus \Omega \subset B_\rho(0) \right\}$$

we have

Theorem 4.7 _For any $p \in (2, 2^*)$ there exists $\bar{\varepsilon}(p)$ such that $\forall \varepsilon \in (0, \bar{\varepsilon}]$ problem (4.1) has at least $\mathrm{cat}_{\bar{\Omega}}[\bar{\Omega}, \mathbb{R}^N \setminus B_\rho(0)] + 1$ distinct solutions. Moreover $\forall \varepsilon > 0$ there exists at least a solution of (4.1)._

We point out briefly the difficulties that to studying (4.1) involves in the case Ω unbounded and the method used to overcoming them.

The solutions of (4.1) are nontrivial critical points, belonging to the positive cone of H_0^1, of the functional

$$J_\varepsilon(u) = \int_\Omega (\varepsilon |\nabla u|^2 + u^2)\, dx$$

constrained on the manifold

$$V = \left\{ u \in H_0^1(\Omega) : |u|_{L^p} = 1 \right\}.$$

The functional J_ε does not satisfy the Palais-Smale condition in all the energy range; the infimum of it on V, $m(\varepsilon, \Omega)$, is just an energy level in which P-S fails and, more, $m(\varepsilon, \Omega)$ is never achieved. So it would be not possible to use the classical theorem 2.4. The abstract tool, that one can reasonably try to use, is, instead, the following

Theorem 4.8 [F] _Let M be a complete Riemannian manifold and $h \in \mathcal{C}^1(M, \mathbb{R})$. Put_

$$h^c = \{u \in M : h(u) \leq c\}$$

$$K^c = \left\{ u \in M : h(u) = c,\ (\nabla h_{|M})(u) = 0 \right\}.$$

Consider $-\infty < a' < a < b < b' < +\infty$ and suppose that h satisfies the Palais-Smale condition on the set $\{u \in M : h(u) \in (a', b')\}$ and $K_a = K_b = \emptyset$. Then

$$\# \left\{ u \in M\ ;\ h(u) \in [a, b],\ (\nabla h_{|M})(u) = 0 \right\} \geq \mathrm{cat}_{h^b}[h^b, h^a].$$

Subsequently one needs to establish a relation between the relative category of suitable sublevel sets of J_ε and $\text{cat}_{\bar\Omega}(\bar\Omega, \mathbb{R}^N \setminus B_\bar\rho(0))$. To carry out this program, first we study the properties of J_ε obtaining the following informations :

i) J_ε satisfies on V the Palais-Smale condition in the energy interval $(m(\varepsilon,\Omega), 2^{1-2/p} m(\varepsilon,\Omega))$ (see [B.C.3]);

ii) a critical point of J_ε corresponding to a solution that changes sign is a function whose energy satisfies the relation $J_\varepsilon(u) \geq 2^{1-2/p} m(\varepsilon,\Omega)$ (see [C.P.1].

Thus, it is convenient to work between $m(\varepsilon,\Omega)$ and $2^{1-2/p} m(\varepsilon,\Omega)$ not only because of compactness reasons, but also because a solution of energy less than $2^{1-2/p} m(\varepsilon,\Omega)$ does not change sign.

Then, defined Ω^+ and Ω^- in the same way as in section 2 we find two energy levels c_1 and c_2 : $m(\varepsilon,\Omega) < c_1 < c_2 < 2^{1-2/p} m(\varepsilon,\Omega)$ and continuous maps

$$\beta : J_\varepsilon^{c_2} \longrightarrow \Omega^+$$

$$\Psi : \Omega^- \longrightarrow J_\varepsilon^{c_2} \cap \{ u \in V : u \geq 0 \} \equiv (J_\varepsilon^{c_2})^+$$

$$(-\Psi) \equiv \Psi_1 : \Omega^- \longrightarrow J_\varepsilon^{c_2} \cap \{ u \in V : u \geq 0 \} \equiv (J_\varepsilon^{c_2})^-$$

such that $\beta \circ \Psi$ and $\beta \circ \Psi_1$ are homotopic to the embedding $j : \Omega^- \longrightarrow \Omega^+$. It is easy to understand that this choice of β and Ψ cannot be done as in section 2, when the case Ω bounded was studied. Indeed, for instance, since Ω is unbounded, if $u \in H_0^1(\Omega)$, $\int_\Omega x |\nabla u|^2 dx$ is not alwais meaningful and, even if we restrict our attention to the subspace of H_0^1 made up by functions having compact support, if β is defined as in section 2 it is not a continuous function.

Nevertheless, it is possible to define properly those maps and then to show

$$\text{cat}_{J_\varepsilon^{c_2}} \left[J_\varepsilon^{c_2}, J_\varepsilon^{c_1} \right] \geq 2 \, \text{cat}_{\bar\Omega} \left[\bar\Omega, \mathbb{R}^N \setminus B_\bar\rho(0) \right].$$

The above inequality, together theorem 4.8, implies the existence of at least 2 $\text{cat}_{\overline{\Omega}} \left[\overline{\Omega}, \mathbb{R}^N \setminus B_{\overline{\rho}}(0) \right]$ solutions that does not change sign and then at least $\text{cat}_{\overline{\Omega}} \left[\overline{\Omega}, \mathbb{R}^N \setminus B_{\overline{\rho}}(0) \right]$ of (4.1).

To obtain the existence of another solution, the idea is to show that, between the levels c_2 and $2^{1-2/p} m(\varepsilon,\Omega)$ there is a change of the topological type of the sublevel sets, that can be explained only by the presence of a critical point. The argument is similar to that described in section 2. In fact we construct a set $\Sigma \subset \overline{\Omega}$ such that $\Psi(\Sigma)$ is not contractible in $J_\varepsilon^{c_2}$ and contractible in a sublevel J_ε^γ corresponding to a higher energy. Hovewer it must be remarked that the proof is very delicate because it involves some careful estimations in order to show that these phenomena occur below $2^{1-2/p} m(\varepsilon,\Omega)$.

References

[A.R.] Ambrosetti A. - Rabinowitz P.: J. Functional Anal. 14 (1973). 349-381.

[Ba-C] Bahri A. - Coron J.M.: Comm. Pure Appl. Math. 41 (1988) 253-294.

[Ba-L] Bahri A. - Lions P.L.: Preprint.

[B] Benci V.: Ann. di Mat. Pura e Appl. (1981).

[B.C.1] Benci V. - Cerami G.: Arch. Rational Mech. Anal. 114 (1991) 13-93.

[B.C.2] Benci V. - Cerami G.: Preprint n.2 (1992) Ist. Mat. Appl. "U.Dini" Pisa.

[B.C.3] Benci V. - Cerami G.: Arch. Rational Mech. Anal. 99 (1987) 283-300.

[Be-L] Berestycki H. - Lions P.L.: Arch. Rational Mech. Anal. 82 (1983) 313-346; 347-376.

[B.N.] Brezis H. - Nirenberg L.: Comm. Pure Appl. Math. 36 (1983) 437-477.

[B.O.] Brezis H. - Oswald L.: Nonlinear Analysis T.M.A. 10 (1986) 55-64.

[B.C.P] Benci V. - Cerami G. - Passaseo D.: "Nonlinear Analysis - A Tribute in honour of Giovanni Prodi". Quaderni Scuola Normale Superiore Pisa (A.Ambrosetti - A.Marino ed.) Pisa (1991) 93-107.

[C] Coron J.M.: C. R. Ac. Sc. Paris 299 séries I (1984) 209-212.

[Ca] Candela A.M.: Preprint.

[Ch] Chang K.C.: Seminaire de Mathematiques Supérieures, 97, Presses de l'Université de Montreal. (1986).

[C.P.1]   Cerami G. - Passaseo D.: <u>Nonlinear Analysis T.M.A.</u> 18 (1992) 103-119.
[C.P.2]   Cerami G. - Passaseo D.: Preprint.
[C.M]   Coffman C.V. - Marcus M.M.: Preprint.
[D.1]   Dancer E.N.: <u>J. Diff. Equations</u> 74 (1988) 120-156.
[D.2]   Dancer E.N.: <u>Bull. London Math. Soc.</u> 20 (1988) 600-602.
[Di]   Ding W.Y.: Preprint.
[E.L]   Esteban M. - Lions P.L.: <u>Proc. Royal Edimbourgh Soc.</u> 93 A (1982) 1-14.
[F]   Fadell E.N.: <u>Raccolta Seminari. Mat. Univ. Calabria</u> n.6 (1985).
[G.N.N]   Gidas B. - Ni W.M. - Nirenberg L.: <u>Comm. Math. Phys.</u> 68 (1973) 209-243.
[L]   Lazzo M.: <u>C. R. Ac. Sciences Paris</u>.
[K]   Kwong M.K.: <u>Arch. Rational Mech. Anal.</u> 105 (1989) 243-266.
[K.W]   Kazdan J. - Warner T.: <u>Comm. Pure Appl. Math.</u> 28 (1975) 567-587.
[N.N.]   Ni W.M. - Nussbaum R.D.: <u>Comm. Pure Appl. Math.</u> 38 (1985).
[P.1]   Passaseo D.: <u>Manuscripta Math.</u> 65 (1989) 147-166.
[P.2]   Passaseo D.: Preprint.
[Po]   Pohozaev S.: <u>Soviet Math. Dokl.</u> 6 (1983) 1408-1411.
[R]   Rey O.: <u>J. Nonlinear Analysis T.M.A.</u> 133 (1989) 1241-1249.
[S]   Stampacchia G.: <u>Seminaire de Mathematique Supérieures, Presses Univ. Montréal</u> 16. (1966).

Index

Banach space 137, 193
Bernoulli shifts 161
Betti numbers 22
Birkoffs ergodicity hypothesis 77
Bochner type inequality 64
Borsuk-Ulam theorem 139
Brezis-Lieb lemma 217

Cartan-Killing inner product 63
Christoffel symbols 2
Colomb gauge 66
Courant-Weyl variational principle 137

de Sitter space-time 226
Dirichlet problems 113, 205
domain shape 263

Ekelands variational principle 193
eliptic periodic solutions 77
Euler-Lagrange equation 2

Fenchel-Legendre transform 80
Floquet multiplier 81

Galerkin approximation 228
Gaussian curvature 55
geodesics 1
Gierer-Meinhardt system 39
Gronwall lemma 71, 156

Hamiltonian system 129, 187
harmonic maps 33
Hessian form 4
Hilbert space 58, 68, 227
Hofner-Zehnder 187
homoclinic orbits 161
Hopf-Rinow theorem 222

Jacobi tangent field 21

Kazdan-Warner problem 105
Keplerian-like potential 119

Laplace-Beltrami operator 103
Laplacian theorem 137
Lebesque measure 78
lightlike geodesics 33
Lipschitz functions 51, 193
Ljusternik-Schnirelman theory 17, 157, 206, 248, 265
Lorentzian manifold 1, 153, 221

mean curvature 39
minimization 49
Minkowski space-time 4
Morse index 2, 55, 60, 85, 228, 265
Mountain Pass theorem 84, 137, 163, 199
multiple pendulum type problem 129

n-body type system 245
Neumann boundary condition 39, 205

Palais-Smale condition 17, 52, 57, 68, 87, 92, 133, 152, 175, 194, 230, 247, 281
periodic solutions 67
Poincare map 78

Riemannian manifold 3, 63, 149, 222, 283

Saddle Point theorem 230
Sards theorem 22, 61
Schwarzchild metric 5
Sobolev space 51, 139, 227, 248, 281
space-time manifolds 1
Stampacchia method 50
static space-time 4
superquadratic growth 83

Uhlenbecks singularity theorem 66

Yang-Mills flow 63